The Weather Surfer

A Guide to Oceanography and Meteorology
For the World's Wave Hunters

By
Vic Morris and Joe Nelson

Grossmont Press　　　　　　　　　San Diego

GROSSMONT PRESS, INC.

7071 Convoy Court   San Diego, California 92111

Copyright ©1977 by Vic Morris and Joe Nelson. All rights reserved. No part of this publication may be reproduced, stored in a retrieval system, or transmitted, in any form or by any means, electronic, mechanical, photocopying, recording, or otherwise, without prior written permission of the author and publisher. Manufactured in the United States of America.

ISBN 0-913182-87-7

Library of Congress Catalog Card Number: 77-71243

Table of Contents

Acknowledgement...vii
About the Authors..ix
Introduction...xi
PART I
WEATHER AND OCEANOGRAPHY FOR SURFERS...............1
Chapter I.   General Weather and Oceanographic Principles
  1. What Causes Waves and Surf..............................3
     Wave generation processes
     Significant wave height
     Wave decay
     Effect of fetch not pointed directly at forecast point
     Travel time of swell
     Wave spectrum
  2. Open Ocean Swell Reaches Shallow Water....................7
     Angled swell approach
     Refraction effects of points and bays
     Description of refraction diagram
     Use of refraction diagram
     Effects of density variations on wave height
     Effect of the continental shelf
     Types of shallow water bottom materials
     Tides
     Local winds and currents
     Presence of more than one swell
     Preliminary comments on surf forecasting
  3. The Major Swell Producers...............................14
     Frontal storms
      Earth's heating system
      Air masses
      Relationship of wind speed to pressure gradient
      Location of frontal storm formation
      Frontal storm life cycle
    Tropical Cyclones......................................19
     Tropical cyclone description
     Causes of tropical cyclones
     Tropical cyclone life cycle
      Initial stages
      Recurvature
      Irregular storm tracks

      Weakening and dissipation
      Interaction with frontal storms
  4. March 1974 Pacific Super Storm............................ 21
      Tropical storm Amy
      Interaction of Amy with Japanese frontal storm
      Generation of high seas aimed towards Hawaii
      Weakening of the storm
  5. Other Sources of Strong Winds............................. 22
      Strong high centers and monsoon winds
      Effect of local land and sea breezes on surf
  6. Ocean Temperatures and Currents.......................... 25
      Global distribution of sea surface temperatures
      Sea temperature and individual comfort
      Distribution and causes of ocean currents
  7. Climate Over the Oceans................................... 32
      Basic circulation features
      Weather associated with circulation features
Chapter II. Forecasting Weather and Surf
  1. Sources of Weather Information in the U.S................... 39
      Marine forecast
      TV and newspapers
      Tropical cyclone warnings
      Local wind forecast
  2. Elementary Forecasting.................................... 40
      Middle and high latitude weather signs
        Weather sequence with a high pressure center
        Weather sequence with a low pressure center
        Weather with cold upper lows
      Tropical weather signs..................................... 45
        Consistency of tropical weather
        Indications of bad weather
        The near equatorial trough
        Monsoon regions
      Forecasting by observing the clouds......................... 46
        High clouds
        Middle clouds
        Convective clouds
        Low clouds
        Other weather signs in the sky
      Swell observation as a forecast aid........................... 48
  3. Basic Swell and Surf Forecasting............................ 50
      Determining fetch area
      Computation of seas in fetch area
      Computation of swell travel time
      Deep water swell and surf height at forecast point
      Sea state near tropical cyclones
      Correction for storm motion

PART II.
REGIONAL SURFING POTENTIAL AROUND THE WORLD......59
Introduction to Regional Sections.................................61
Chapter I. North Atlantic
  1. North Atlantic Wind Circulation...........................63
     The Bermuda-Azores high
     Frontal storm surf
      North America East Coast
      Bahamas, Antilles, N.E. South America
      Western Europe and N.W. Africa
     Frontal storms in warmer months
  2. North Atlantic Hurricanes................................67
     Early season storms
     August tropical cyclones
     September tropical cyclones
     October tropical cyclones
     Late season storms
     Variations in hurricane frequency
  3. Other Sources of North Atlantic Surf........................70
     The Bermuda high
     Carribbean west winds
  4. Other Important Factors in Evaluating Surf Potential...........71
     Local wind conditions
     Bottom conditions
      The continental shelf
      Shallow water near shore
Chapter II. South Atlantic
  1. South Atlantic Wind Circulation...........................73
     Lack of tropical cyclones
  2. "Roaring Forties" Frontal Storms..........................73
  3. Geographical Effects on South Atlantic Surf...................75
     Low latitude eastern South America
     Middle latitude eastern South America
     Southwest Africa
     Equatorial West Africa

Chapter III. Indian Ocean
  1. The Monsoon...79
     Northern Hemisphere summer
     Northern Hemisphere winter
     Transition seasons
  2. Tropical Cyclones.....................................80
     Northern Hemisphere
     Southern Hemisphere
  3. Indian Ocean Surfing Potential............................82
     Arabian Sea and Bay of Bengal coastlines
     East coast of South Africa

      South Indian Ocean islands
      West Australia
        North coast
        Northwest coast
        Central west and southwest coast
      South Australia
Chapter IV.   The Pacific
  1. North Pacific Circulation............................87
      Summer
      Fall
      Winter
      Spring
  2. North Pacific Tropical Cyclones........................89
      West North Pacific
      East North Pacific
  3. South Pacific Circulation............................91
      Frontal storms
      Tropical cyclones
  4. North Pacific Regional Surf...........................96
      South China Sea
      Japan
      Western North Pacific islands
      Hawaii
        North Shore
        South Shore
      Pacific Northwest
      Central California
      Southern California
      Baja California
      South Mexico and Central America
  5. South Pacific Regional Surf...........................101
      Western South America
      Central South Pacific islands
      New Zealand
      Northeast Australia
      Southeast Australia
In Conclusion...102

Acknowledgements

The authors wish to thank the following organizations and people. They contributed time, materials, and most of all the inspiration essential for the preparation of this book: National Weather Service Forecast Office, Honolulu, Hawaii; *Surfer Magazine; Surfing Magazine;* Steve Mangiagli Manufacturing Company; Honolulu Surfboards; Glen Matayoshi; Cheryl Coambs; and Hector Arechiga.

V.M.
J.N.

About the Authors

Separately and together, Vic Morris and Joe Nelson have been surfing since they were old enough to carry a surfboard—Baja, Hawaii, Australia, New Zealand, Central America, and up and down the American coasts. Between them they've logged over 20 years in search of the big ones.

The two first met in California in 1972. It was an auspicious get-together. According to his co-author, Joe was the one who asked the questions that got THE WEATHER SURFER written. A long-time resident of Southern California (he grew up around Hermosa Beach), Joe has spent most of his life doing what the rest of us would give our eye teeth to do—following the waves.

Vic, "the weatherman," lived in Hawaii for five years and put in several summers as a lifeguard on Cape Cod. He comes by his know-how naturally; he received his M.S. in meteorology from the University of Hawaii in 1975, and spent two years as a weather forecaster for the U.S. Navy fleet in Pearl Harbor.

The authors currently live in Cardiff-by-the-Sea, California, where Joe works as a custom home contractor and Vic teaches geography at San Diego Mesa College.

Introduction

In his mind's eye every surfer has an image of the perfect wave. It may be a local beach break on a crisp, offshore autumn day. Or it may be crashing, cobalt cylinders breaking with awesome force on an unexplored tropic reef.

But whatever type of surf he seeks, there are two unpleasant realities in a surfer's life. Surfing is a growing sport with ever increasing numbers competing for available waves. And growth of coastal communities almost inevitably leads to decreased access to surfable waves. With these pressures building many surfers are forced to travel greater and greater distances in search for good uncrowded surf.

A major surf trip requires a substantial investment in planning, travel time, and often money. Experienced traveling surfers are usually well prepared. They have the basic living necessities, a suitable car or jeep, and a full surfboard repair kit. But few of them have the simple knowledge of oceanography and meteorology that could mean many more waves to themselves. Yet there are still more than enough excellent and uncrowded waves left in the world to reward those who search for them.

As every surfer knows, waves are a very passing phenomenon. Swells build and die, winds constantly change direction, sometimes in a matter of hours or minutes, greatly changing the quality of the surf. But many of these changes follow fairly regular and understandable patterns that are worldwide. The basic knowledge of oceanography and meteorology presented in this book will be valuable in planning future surf trips, to the next town or halfway around the world.

In Part I of this book we discuss the processes that produce waves in the open ocean. Then we see what happens when they reach shallow water. This is where you will learn about wave refraction and the types of coastal features that produce good surf.

Next we consider some basic meteorology and discuss the two primary swell producers: the frontal storm and the tropical cyclone. To illustrate the principles we do a follow-up on the March 1974 super storm that produced 40-foot waves in Hawaii. Following this are discussions of other causes of strong winds, descriptions of ocean temperatures and currents, and a section on climate over the oceans.

The next section describes how to do your own weather forecasting. For the United States the many sources of weather information are described. Other forecasting methods such as observing the clouds, barometer reading, local winds, and swells are also discussed. With this information you will have a good idea what conditions the next day will bring.

Part II uses the principles discussed in Part I to evaluate the surfing potential of the major ocean areas. There are separate discussions of the wind circulation features and surf conditions in four major regions: North Atlantic, South Atlantic, the Indian Ocean and the Pacific Ocean.

PART I

Weather and Oceanography For Surfers

Chapter I

General Weather and Oceanographic Principles

1. What Causes Waves and Surf

Wave Generation Processes

Waves are produced by the wind blowing over the sea surface. The height of the waves produced depends upon three factors: speed of the wind, the length of time the wind has been blowing (duration), and the fetch length. The fetch is the distance over which the wind has been blowing at a relatively steady speed and constant direction uninterrupted by solid land masses.

Typical fetches of wind in strong winter storms are 500 to 1,500 miles long and usually several hundred miles wide. For each combination of speed, fetch, and duration, waves of a specific average height and average period are produced. The period is the time interval between the passage of two wave crests at the same point. Waves produced in the fetch continue to grow until they reach a steady state known as a fully developed sea. Until this point an increase in the wind speed, fetch, or duration will cause the swell to become larger and have a longer period.

Significant Wave Height

Oceanographers often refer to the significant wave height. This is the mean height of the highest one-third of the well-defined waves observed at a given point. The significant wave usually equals one and one-half times the average wave height. In surfing this would correspond to the bigger set waves. Mathematically it has been shown that the highest wave out of 1,000 will be nearly twice the significant height or almost three times the average wave height. In a day's surfing, you may very well see 1,000 waves break on your favorite reef. So while surfing a 6-foot average (9-foot set waves) at Sunset Beach, Hawaii, there theoretically may be an 18-foot cleanup wave waiting to crush the unwary. It's something to think about!

General Weather and Oceanographic Principles

Wave Decay

(Distance and Wave Period Effects)

As soon as the waves leave the area in which they were generated (fetch) they begin to decrease in size. The percentage of height decrease depends primarily on the distance the waves travel and to a lesser degree upon their period.

The longer period (and usually larger) waves transmit most of their energy directly downwind. They are less steep than short period waves and do not decrease much due to wind resistance.

Shorter period waves tend to spread considerable energy at angles up to 90 degrees from the direction of the wind producing them. The slope of a short period wave is relatively steep and its height can be reduced considerably by opposing winds. (Fig. 1)

The height decrease due to decay is at first quite rapid and then it occurs more slowly with time. Let's consider a couple of examples to illustrate the points discussed in the preceding paragraphs. (Fig. 2)

		Case A	Case B	Case C
Initial Conditions: Fetch Width Assumed ½ Fetch Length	Wind Speed	35 kts	60 kts	35 kts
	Duration	40 hrs	6 hrs	6 hrs
	Fetch Length	650 nm	100 nm	100 nm
	Highest ⅓ Height	25'	25'	11'
	Mean Period of Highest ⅓ Waves	14.5 sec	11.5 sec	8 sec
500 Mile Decay Distance	Highest ⅓ Height	12'	6'	2'
	% of Initial Height	.47	.24	.22
1,000 Mile Decay Distance	Highest ⅓ Height	9'	4'	1½
	% of Initial Height	.36	.16	.13
2,000 Mile Decay Distance	Highest ⅓ Height	7'	2½'	1'
	% of Initial Height	.27	.11	.09

fig. 1

It can easily be seen what the effect of wave period is in cases A and B. Case A's 25-foot waves are typical for those produced by a moderate strength, slow-moving, large winter frontal storm. After 2,000 miles of decay the waves still have 27% of their initial height. The resultant 7-foot open water swell could produce surf of 10 feet or more at a well-exposed coastal point. (The relationship of deep open water swell height to surf height is discussed in the sections on wave refraction.)

In case B the near hurricane force winds of 60 knots raise a 25-foot sea in 6 hours, but the mean period of the waves is less than in case A. By the time these waves travel 2,000 miles, only 11% of their initial height remains. The 2½-foot open water swell probably will produce only 3-foot to 4-foot surf when it reaches shore.

Case C. illustrates the effect of limiting fetch length and duration of a

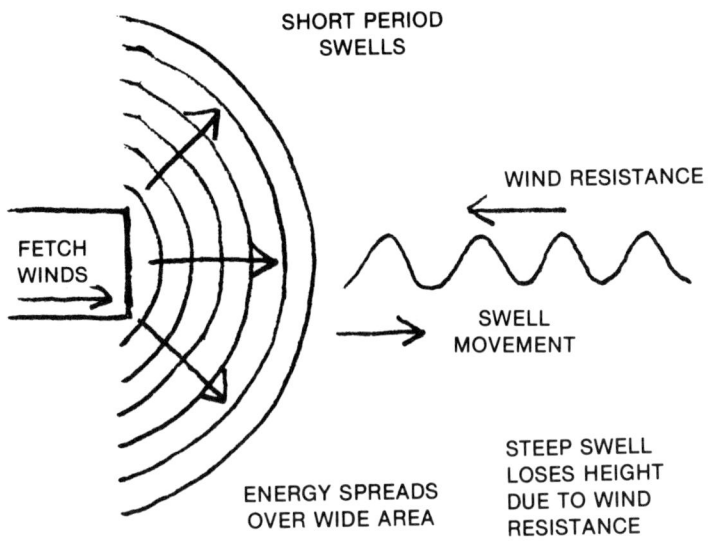

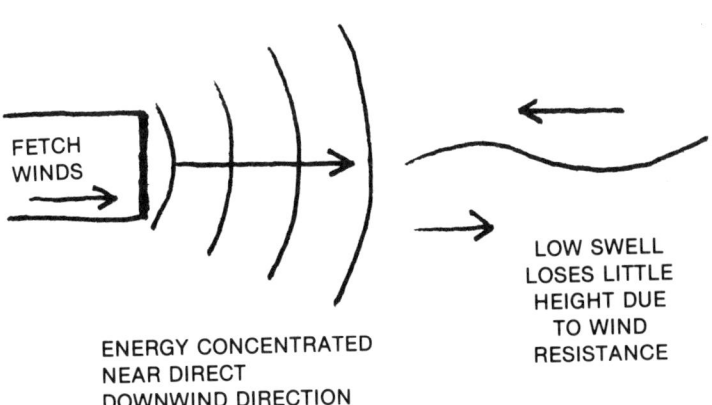

fig. 2

General Weather and Oceanographic Principles

blow. The 35-knot winds are the same speed as in case A but the resulting initial seas are only 11 feet. Only 9% of the initial height is left after a 2,000 mile journey, leaving a barely noticeable 1-foot swell.

This decayed swell height will be the final deep water swell arriving at a point only if the swells have been traveling along a straight line from their source to their destination (winds in the fetch aimed directly at point of interest), and if they have not encountered strong winds or sizable waves in the decay area. These effects could act to either reinforce or reduce the traveling swell depending on their direction.

Effect of Fetch Not Pointed Directly at Forecast Point

If the generating fetch is not pointed directly towards the forecast point only a percentage of the expected swell height will reach the point in question. Quite obviously, the greater the angle between the wind direction in the fetch and a line from the center of the forward edge to the forecast point, the smaller the waves will be at the destination.

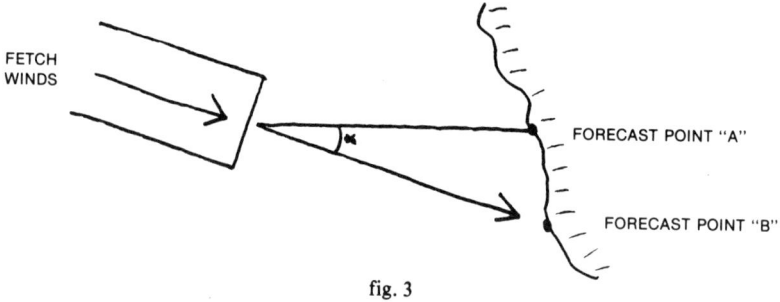

fig. 3

For practical purposes if the angle * indicated on the diagram is 15 degrees or less, most of the decayed swell generated in the fetch will reach the forecast point.

It's time for a note of caution. On most maps a straight line does not represent the shortest distance between two points. The error is not too important for distances up to 1500 miles. But for greater distances one must determine the great circle path the waves are following.

A great circle is the shortest distance between two points on a sphere. Special map projections can be used to plot great circle paths.

Travel Time of Swell

The time a swell takes to travel from its source to any point depends upon the wave period. The swell speed in knots is approximately 1.5 times its period in seconds. (See example at end of wave spectrum discussion.)

General Weather and Oceanographic Principles

Wave Spectrum

Seas produced within a fetch have a range of periods deviating from a mean value that is determined by a given set of wind speed, duration, and fetch length conditions. This feature is known as the wave spectrum. Waves having a period near the mean of the spectrum possess the greatest energy and are largest.

The longest period waves travel more rapidly, and the shorter period ones are left behind. This process is known as dispersion. As the swell moves away from the generating fetch, the period of the waves as observed along a line towards their destination becomes more distinct. This causes "cleaner," more well-defined swell waves, the kind that surfers want.

At a point several hundred miles or more downwind from a fetch the following sequence of events is usually observed: First to arrive are long period (often 15 to 20 seconds or more) waves containing relatively little energy, which are quite small. But to a sharp eye these are signs of a building swell. As the slightly slower, somewhat shorter period waves arrive, the swell builds to maximum height. These medium period (often 8 to 18-second) waves contain the most energy. Gradually the wave period shortens further and the wave height decreases as swells of lesser energy arrive. Often the really short period waves (5 seconds or less) never travel more than a few hundred miles beyond the fetch before being flattened out by other prevailing winds or seas in the decay area.

The duration of a swell at a point depends mostly on the length of time the wind has blown through the fetch, and the distance over which wave dispersion has occurred. Surf produced by swells arriving from great distances usually lasts longer than surf from nearby swell sources.

Suppose there are swells with periods ranging from 10 to 20 seconds arriving from 1,000 miles away. The 20-second swell will arrive at the forecast point in 33 hours, while the 10-second swell will arrive in 66 hours. Waves will last for 33 hours.

Now let's say the same period swells travel for 3,000 miles. The 20-second period swell will arrive in 100 hours, while the 10-second period swell will arrive in 200 hours. In this case, the swells from the 3,000 mile journey last 67 hours longer than those traveling only 1,000 miles. However the swell traveling 3,000 miles will not be too large unless the initial wind waves are huge.

2. Open Ocean Swell Reaches Shallow Water

When the open ocean swell reaches shallower water near shore, it steepens and finally breaks into surf. Several factors influence the amount of growth that occurs from a deep water swell to a breaking

General Weather and Oceanographic Principles

photo: Paul Heussenstamm/SURFING MAGAZINE

wave. These include the angle of approach the deep water swell makes with shore, the configuration of shallow water bottom contours, the slope of beach, and the deep water steepness of the wave (the ratio of wave height to wave length). The greater growth occurs when swells having little steepness and long period approach a strongly sloping beach. In areas of gently sloping bottom contours, waves of average steepness break at a water depth of about 1.3 times their height. (Fig. 4)

Angled Swell Approach

Swells approaching from some angle will grow less than those approaching perpendicularly to the beach. As the angled deep water swell approaches, shore refraction processes cause the swell in shallow water to approach the beach more nearly head-on.

Refraction Effects of Points and Bays

As waves enter shallow water (oceanographers define "shallow" as equal to ½ a wave length), they begin to feel the effect of bottom configurations. As the water becomes shallower the waves slow down and steepen.

If the bottom contours are non-parallel, as in the case of points and bays, different parts of the wave front will be traveling over varying depths of water. The wave front in deeper water (a bay) travels faster than the part in the shallower water (near a point). This causes the

STRONGLY SLOPING BOTTOM CONTOURS
AND LOW SWELL STEEPNESS
(LONG PERIOD) YIELD MAXIMUM
WAVE GROWTH

SHALLOW SLOPING BOTTOM CONTOURS
AND HIGH SWELL STEEPNESS
(SHORT PERIOD) YIELD MINIMUM
WAVE GROWTH

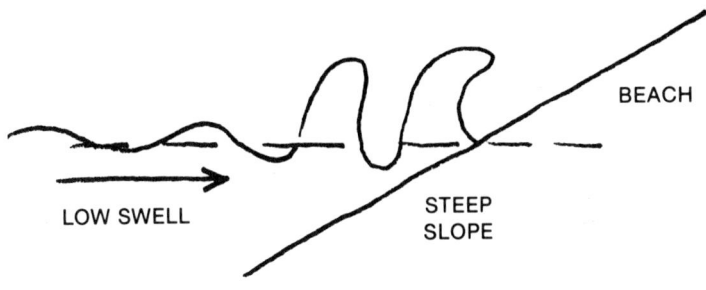

fig. 4

General Weather and Oceanographic Principles

wave fronts to roughly parallel the bottom contours near shore. The wave energy is concentrated near the point and diffused over a wider area in the bay. This explains the fact that the largest surf breaks near pronounced points of land. (Fig. 5)

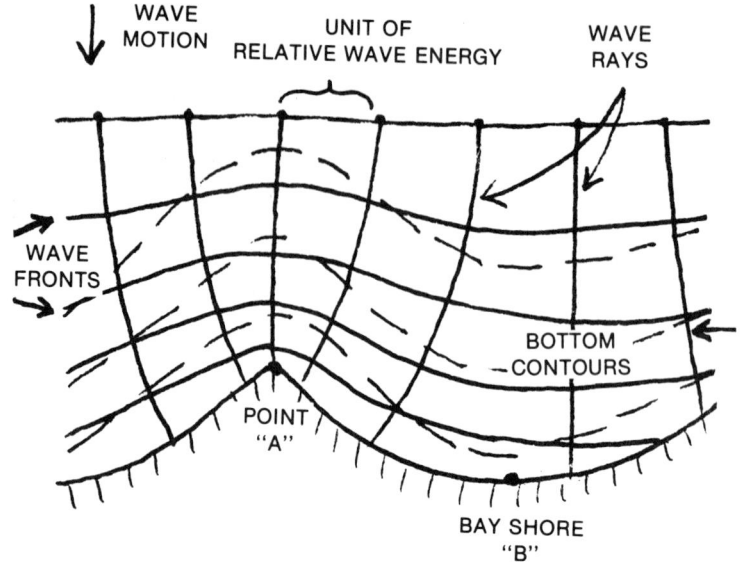

REFRACTION DIAGRAM

IN SHALLOW WATER WAVE SPEED
IS PROPORTIONAL TO THE SQUARE
ROOT OF THE WATER DEPTH

fig. 5

Description of Refraction Diagram

The accompanying illustration is a refraction diagram. It shows the bending of the wave fronts produced by changes in the depth of the shallow water contours. The position of the wave front is shown at equal time intervals. These positions are calculated from the formula for shallow water wave speed. Next, the deep water wave front is divided into equal parts, and wave rays are drawn perpendicular to each wave front. The horizontal distance between two wave rays is a measure of the relative wave energy present. When wave rays become close together a lot of wave energy is present, and there will be large growth from deep water swell height to breaking wave height.

General Weather and Oceanographic Principles

Use of Refraction Diagram

By applying these principles, one can make an estimate of bottom contour effects for various wave directions, heights, and periods. Sometimes offshore canyons and ridges can focus wave energy sufficiently to cause surf height at one particular beach to be several times larger than at nearby locations. In the case of parallel bottom contours (straight beach) and a swell directly approaching shore, surf forecasters assume the surf height will be 1.5 to 2.0 times the deep water wave height.

Effects of Density Variations on Wave Height

The computations for the wave height assume ocean water of relatively constant density. However, whenever this is not true near shore, the higher density conditions may alter the expected surf heights. Near some rivers or estuaries, mud or plankton may increase the density of the water enough to reduce expected wave heights.

Areas of high kelp concentration damp out short period wind chop to a large degree, but allow longer period swells to pass through with only slight decay. This desirable feature allows comparatively glassy conditions even when moderate onshore winds are prevailing.

Effect of the Continental Shelf

Once a wave reaches the "shallow water" depth of one half a wavelength, it begins to feel effects of the ocean bottom and its forward speed slows. (One wave length in feet $= 5.12T^2$ where T is the wave period in seconds.) Waves traveling over long expanses of relatively shallow water will be slower in forward speed than those initially reaching shallow water near shore. This explains the fact that waves of a given size usually move faster in Hawaii than on the mainland U.S. shorelines.

Types of Shallow Water Bottom Materials

As previously noted, the type of breaker produced depends largely upon the slope of bottom contours near shore. Shallow water coral and rock bottoms are firmly anchored and quite often have steep slopes, resulting in hard breaking waves.

Sand bottoms consistently move due to currents and surf. From one day to the next the best place to surf may change drastically. The seaward edge of sandbars usually slope gradually. This often results in less violent spilling breakers that may tend to become "mushy."

One thing to keep in mind when traveling to tropical regions is that

General Weather and Oceanographic Principles

coral will not grow if the salinity or temperature of the water is reduced by a cool fresh water stream. This will form a break in a barrier reef and a chance for good surf.

Tides

Most surf spots seem to break best at a preferred water depth that varies according to surf height. In some areas of small tidal range, such as Hawaii, the 2-foot variation in water level is not too important. In some sections of New England which have 10-foot or greater tides, the time of tide is almost everything. It can mean the difference between unbroken swells hitting the beach, closed out conditions over continuous sandbars, or good surf.

Knowledge of tides is very important for a traveling surfer.

Local Winds and Currents

The effect of local wind direction on waves is well known by any regular surfer. Offshore winds prevent a wave from breaking until it is steep and hollow. Onshores cause a breaker to collapse early, producing a spilling, mush wave. Furthermore, the wave faces will be marred by wind chop produced by the onshore wind.

If you happen to surf at a place exposed to a seaward moving current or a rip, the waves will steepen and break in unusually deep water. Shoreward moving currents reduce wave height much as an onshore wind does. In either case, a strong current will cause choppy, bumpy waves. Currents near shore can be produced by tides, winds, or fluctuations in sea level due to surges of white water.

Presence of More Than One Swell

So far we've considered only surf produced by one swell present at a time. At times, especially during the winter storm season, two or more major swell trains may be present simultaneously at a point. This is particularly apt to be true for islands in mid-ocean.

If two or more swells arrive simultaneously that produce surf, complex interactions can occur. To determine the resulting wave height, one must account for the size, direction, and period of each of the contributing swells. Whenever the wave trains of the swells are in phase with each other (the wave crests of both swells arrive at the same time), the resulting wave height is the same as the sum of contributing wave heights. When they are out of phase (the wave crests of one swell arrive at the same time as the trough of the other swell), the final wave height is the difference between the individual heights. The wave trains

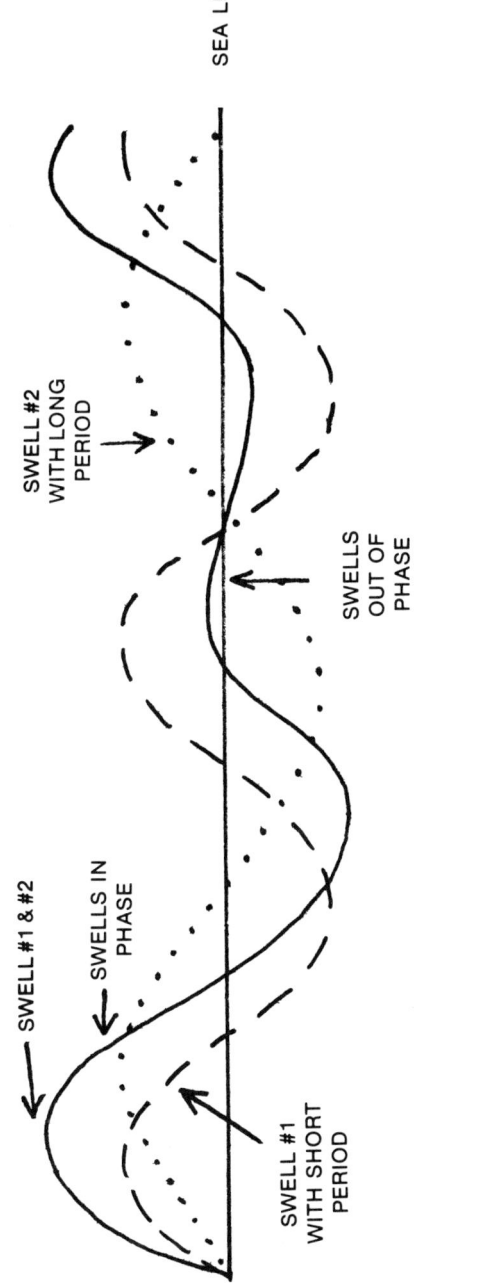

fig. 6

General Weather and Oceanographic Principles

from the same direction will be in phase at intervals equal to the individual wave periods multiplied by each other. (Fig. 6)

For example, suppose one swell has a 10-second period and a second swell has a 15-second period. They will be in phase at a 10 x 15 = 150 second or a 2½-minute interval.

For the surfer, the waves on mixed swell days (if two or more swells are sizable compared to each other) will be irregular. Over a several minute period or more, the waves of each successive set will increase in size. Then they will reach a peak size and afterwards decrease. Long lulls with few waves will occur when the swells are out of phase with each other. If one of the swells has a much shorter period than the other, the surf is apt to be sloppy and difficult to line up in.

Preliminary Comments on Surf Forecasting

By now the reader can see that predicting surf accurately is a very complicated process requiring numerous diagrams and several days of accurate weather and wave observations over a vast expanse of ocean.[1] Fortunately, a basic knowledge of the principles involved is enough for a surfer to figure out when surfable waves are likely to be at whatever spot he might be interested in.

3. The Major Swell Producers

As has already been noted, waves are produced by wind blowing over a given fetch for some length of time. In order to determine whether or not surf is likely at a given spot, one needs to know the possible sources of swells that can reach that location. Large swells (which are the easiest to observe, track, and predict) are produced by organized regions of strong winds that are usually associated with intense frontal storms, or tropical storms and hurricanes. Let's consider some of the principles of basic meteorology involved in understanding the causes and behavior of surf-producing storms.

Frontal Storms

Earth's Heating System

The earth's surface is heated unequally by the sun's rays. Averaged over the year more heat is gained from the sun than is lost to space equatorwards of 35° to 40° latitude. Polewards of these latitudes the cooling by radiation to space exceeds the heating from the sun. Yet

---

[1] Detailed techniques are available from U.S. Hydrographic Office Publication No. 603: "Practical Methods for Observing and Forecasting Ocean Waves by Means of Wave Spectra and Statistics." R.W. James, G. Neumann, and A.J. Pierson, Washington, D.C., 1960.

General Weather and Oceanographic Principles

year after year there is little net change in the mean annual temperature anywhere on the earth.

Warm air is less dense than cold air. Whenever unequal heating takes place the colder, denser air moves to replace and undercut adjacent warm air. This causes the wind systems of the earth, which distribute the atmosphere's heat and prevent major long-term temperature changes.

Air Masses

Generally the air masses of the world retain many of the characteristics of their source region despite long journeys. In winter cold air from Canada can cause killing frosts in Florida and California. In the summer, hot humid air can plague the New England states with conditions much like those of an Amazon jungle. The region of transition between two adjacent air masses is known as a front. Frequently fronts are associated with bad weather and often are the spawning grounds of low pressure centers.

Relationship of Wind Speed to Pressure Gradient

The strength of the wind is directly related to the spacing of isobars (lines of equal air pressure). The winds blow nearly parallel to the isobars and slightly across them towards lower pressure. Tightly spaced isobars surrounding intense storms are associated with the highest wind speeds. Strongly developed higher latitude oceanic storms often contain regions of 50 to 60 knot winds blowing over sufficient duration and fetch to generate average seas of 30 to 40 feet or more.

Location for Frontal Storm Formation

Regions of sharp temperature difference over a short distance are the preferred formation areas for strong frontal low pressure centers. In the winter hemisphere the temperature decreases poleward more rapidly than in the summer hemisphere. Therefore, most strong frontal storms occur in the winter hemisphere.

During the winter the higher latitude continental areas are snow and ice covered and become vast reservoirs of extremely cold and dry air. Warm ocean currents move northward off the east coasts of the continents heating the air overlying them. As most mid and high latitude air masses move from west to east, in winter cold Arctic air often reaches the warm ocean water. The most intense fronts of the Northern Hemisphere winter are found off Japan and the U.S. Atlantic coast. These are often the spawning grounds of fast-moving, rapidly intensifying storms that may eventually dominate half an ocean's

General Weather and Oceanographic Principles

weather for a week or more.

In summer the sun warms the mid-latitude regions nearly as much as the tropics. Only higher latitudes remain fairly cool. Therefore, in summer most fronts are weaker and in higher latitudes during the winter. Most frontal storms in summer generally confine their influence to relatively high latitudes. The characteristics of spring and fall frontal storms are usually intermediate between those of winter and summer.

Frontal Storm Life Cycle

Frontal storms follow a regular life cycle. In the predisturbance stage near a stationary front, the air on the colder side of the front is moving 180° opposite to the warm air on the other side. Then at some point a pressure fall starts along the front and a low pressure area is born. The air begins to spiral cyclonically (counterclockwise in the Northern Hemisphere and clockwise in the Southern Hemisphere) around and somewhat towards the center of lowest pressure.

The colder, heavier air begins to move, wedging under the warmer air ahead of it. The boundary is known as a cold front. Further east warm air begins to ride over cold air ahead of the storm, forming a warm front. The intensifying storm receives its energy from the contrast between the warm and cold air masses. Motion of the storm is determined by the winds overlying it.

Most frontal storms move towards a direction somewhere between east and north in the Northern Hemisphere, and towards east, southeast, or south in the Southern Hemisphere.

As the storm further intensifies, the colder air begins to overtake the warm air faster than the warm air recedes. Eventually the cold front catches up with the warm front, forcing the warm air aloft. The result is known as an occluded front. Prior to the occlusion the storm is known as an open wave cyclone, and the lowest pressure is found at the junction of the warm and cold fronts. After occlusion occurs, lowest pressure is located some distance west from the occluded front.

Frontal cyclones attain peak intensity shortly after occlusion occurs. Once the warm air is driven aloft the temperature contrast that maintains the storm disappears. Gradually the air pressure at the low center rises and the winds weaken. Usually weakening occluded low centers have weak upper winds over them and therefore become nearly stationary. (Fig. 7)

TYPICAL FRONTAL STORM LIFE CYCLE

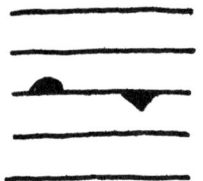

STATIONARY FRONT

FRONTAL WAVE

DEEPENING
FRONTAL WAVE

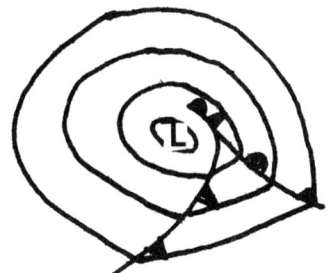

OCCLUDING
AND DEEPENING
LOW CENTER

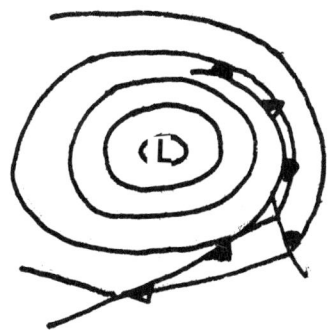

MATURE
OCCLUDED
LOW
CENTER

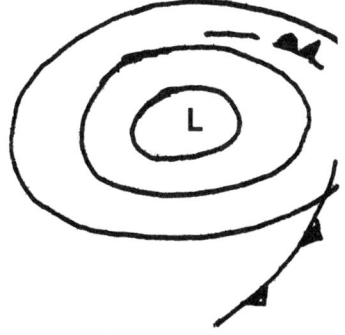

DISSIPATING
OCCLUDED
LOW
CENTER

REFER TO LEGEND FOR SURFACE
ANALYSIS (p. 23) FOR EXPLANATION
OF SYMBOLS.

fig. 7

General Weather and Oceanographic Principles

Tropical Cyclones

Tropical Cyclone Description

The other major swell producer in the world's oceans is the tropical cyclone or hurricane. They are generally smaller than most frontal storms but often much more intense. Sustained maximum winds in intense hurricanes have been estimated at 150 knots or more. Fortunately such extreme winds only happen in a ring a few miles wide which surrounds the calm eye of the storm. Hurricane force winds in severe storms may extend anywhere from 10 to perhaps 50 to 100 miles from the center. A 50 to 300 mile wide doughnut of gale force wind usually surrounds the hurricane force wind zone.

Often the wind pattern around a tropical storm is asymmetrical. Usually higher winds blow over a longer fetch and for a greater duration to the right of a moving storm (in the Northern Hemisphere) than to its left. This results in higher seas being produced in the storm's right semicircle (or "dangerous semicircle," as mariners call it). Generally these seas move parallel to the cyclone's path.

Heavy swells can propagate a thousand miles or more ahead of a slow-moving long-lived hurricane. Usually swell size increases slowly with the storm's approach. However, a fast-moving storm may move nearly as fast as the waves it produces. Swells from a fast-moving storm will build rapidly as it nears and decrease rapidly as it passes by.

Many people do not understand the full meaning of the following terms contained in tropical cyclone warnings. A tropical cyclone is a warm-cored, non-frontal low pressure center of any intensity in tropical or adjacent mid-latitude areas. Tropical depressions are tropical low centers having maximum sustained winds below gale force (33 knots). Low centers with peak winds of gale (34 to 47 knots) or storm (48 to 64 knots) intensity are called tropical storms. Those attaining full hurricane intensity (65 knots) are known as hurricanes, or typhoons in the western Pacific.

The passage of a fully-developed hurricane is an experience not easily forgotten. Intense hurricane winds can flatten all but the most soundly constructed buildings. Objects picked up by the raging winds become lethal missiles. Very heavy rainfall accompanies most hurricanes, often producing rampaging flash floods. Onshore blowing hurricane winds can cause storm surges, raising the ocean level 10 to 15 feet or more higher than the usual tidal range. Hurricane driven seas have nearly wiped some low-lying coastal towns off the map. In winds over 85 knots it is difficult to tell where the sea ends and the air begins.

General Weather and Oceanographic Principles

Causes of Tropical Cyclones

These furies of the seas are born over tropical waters which have temperatures of 80° F or higher. The warm ocean provides the source of heat and moisture required to maintain a hurricane. A number of complex conditions also must be present simultaneously for hurricane development. Of the many areas of bad weather in the tropical oceans, only 1% or less eventually become hurricanes.

Perhaps 60 to 100 hurricanes or tropical storms develop annually worldwide. All of the world's tropical ocean areas except the South Atlantic and portions of the eastern South Pacific occasionally spawn hurricanes. With the exception of those storms developing in the Bay of Bengal and Arabian Sea, the principal hurricane season is the summer and early fall. During these months the sea surface temperatures are usually highest and other meteorological conditions are most favorable for hurricane formation. Within the hurricane season, developments often occur in streaks. Sometimes several storms form in a few days followed by a quiet period of little storm activity for a month or more. The North Pacific is especially noted for sporadic frequency of storm formation.

Tropical Cyclone Life Cycle

Initial Stages

Frequently weather satellites detect the initial formation of a tropical cyclone. Usually several days are required for the storm to attain full intensity, although very sudden intensifications of up to 100 knots in 24 hours or less have been known to take place. Most tropical cyclones form initially between latitudes of 5° and 20°. However some Atlantic and western Pacific storms have formed poleward of 30° latitude over warm ocean currents. During the earlier days of their life cycle, a large percentage of storms are steered westward by trade winds equatorwards of a subtropical high pressure belt. In the Northern Hemisphere, young storms frequently follow a west to northwest course at 5 to 15 knots.

Recurvature

As the storms reach higher latitudes, they usually become influenced by upper level westerly flow from the middle latitudes. If the storm is still going at this point, it will change course sharply. (Many eastern Pacific storms fail to recurve because cold dry surface air from the cool ocean destroys them before they reach recurvature latitudes.) In the

General Weather and Oceanographic Principles

Northern Hemisphere the storm usually assumes a northeast or easterly course, often accelerating to forward speeds of 20 to 40 knots or more. Normally after recurving, the storm slowly weakens. But its circulation may expand, still permitting waves to form as large as those developed at peak hurricane intensity.

Irregular Storm Tracks

The preceding is only a rough mean life cycle that many tropical cyclones follow. These storms sometimes follow very erratic tracks, describing loops with sudden accelerations and decelerations, enough to baffle even the most experienced forecaster.

Weakening and Dissipation

Once a tropical cyclone reaches hurricane intensity, the chances are excellent it will remain intense as long as it continues over warm (over 80° F) ocean waters and no cold, dry air enters the circulation. Only about one typhoon per season (out of about 20) dissipates over tropical waters in the western Pacific. Storm weakening over warm ocean waters is usually due to unfavorable upper air conditions.

Typical tropical cyclones last several days to 2 weeks in the course of their ocean voyages. One storm, Hurricane Ginger, was on Atlantic weather maps for 31 days in 1971 and was producing waves for most of that time.

There are reasons for the death of most tropical cyclones. If the storm moves onshore over a sizable land mass (especially if it is mountainous), the lower level wind circulation will nearly vanish in 24 to 48 hours. However, the upper level remnants might be able to regenerate a storm if the circulation once again moves offshore to warm waters.

Should the storm enter middle latitudes, it will eventually encounter regions of cooler surface water. Since the cooler waters no longer can supply as much heat and moisture energy as tropical waters, the storm must weaken. Furthermore, cold dry air from adjacent polar air masses may be drawn into the storm system, causing further weakening.

Interaction With Frontal Storms

In some cases, however, a tropical cyclone in higher latitudes may be drawn into the circulation of a frontal storm. When this occurs, the cyclonic energy of the tropical cyclone and its heat and moisture is added to the frontal cyclone. This can result in the rapid development of a very dangerous frontal storm. An excellent example of this is given in our case study of the late March 1974 North Pacific storm system.

General Weather and Oceanographic Principles

Such systems can be more potent wave producers than the initial tropical storm or hurricane which fed them.

4. March 1974 Pacific Super Storm

Let's look at an example to see how a storm system can generate surf. In late March 1974, one of the biggest swells in recent years struck the north and west facing coastlines of the Hawaiian Islands. The waves that reached 40 feet along the "country" cloud breaks on Sunday, March 24, were the product of storm energy that had been gathering in the Pacific for over two weeks.

Tropical Storm Amy

As early as March 8, weather conditions became disturbed in the tropical western Pacific, and a weak cyclonic circulation was detected near 5°N, 152°E. For the next several days, the tropical weather southeast of Guam remained showery and squally, awaiting a mechanism to concentrate the energy into a storm. On March 13, a definite low pressure center and increasing cyclonic winds marked the birth of a tropical depression 300 miles south southeast of Guam. This depression drifted northwest, slowly gaining strength from the warm ocean and moisture-laden tropical air.

On the 16th, Tropical Storm Amy was born 350 miles west northwest of Guam, packing gale force 45-knot winds near its center. Up to this point, the sequence of events was typical for a developing tropical storm. Usually an out-of-season tropical cyclone cannot spread its effects far from the tropics before being destroyed by cold air. But, Amy's demise was to fuel a monster a hundred times larger and far more intense than herself.

Interaction of Amy with Japanese Frontal Storm

On the 16th and 17th a typical winter storm took shape off Japan and began to drift eastward. The next day Amy felt the influence of the storm and accelerated to the northeast. The Japanese storm started to intensify and would undoubtedly become a potent wave maker as many do during this season in the North Pacific.

By March 20, Tropical Storm Amy's attempt to outrace the expanding Japanese storm met with failure and cold air entered her circulation. Energy of the dying tropical storm became absorbed into the frontal storm. The result of this marriage was a sudden 20 to 30 millibar deeping of the combined storm system along with the establishment of a 1,000 mile long 40 to 60 knot fetch of northwesterly wind reaching almost to Midway Island.

General Weather and Oceanographic Principles

Generation of High Seas Aimed Towards Hawaii

On the 21st the Pacific colossus crossed the Dateline near 40°N, continuing on a steady easterly course. The central pressure of the storm was around 950 millibars (more than 60 mbs. below normal sea level pressure). Maximum sustained winds were now near hurricane force of 65 knots. The gargantuan fetch of 40 to 60-knot winds now extended almost from Japan to a few hundred miles from Hawaii. Anyone with a trace of weather wisdom could see that the Sandwich Isles were in for trouble, big, big surf trouble. Repeated high surf warnings were broadcast to the public beginning late on the 21st.

And as if this weren't enough, a new storm moved eastward off Japan on the 22nd to join its forces with the main storm northwest of Hawaii. On the 23rd the stage was set for monster surf as the huge fetch extended almost to the shores of Kauai. The central Pacific storm became nearly stationary 1,000 miles from Honolulu.

Fortunately for Hawaii, through the day the main fetch became oriented more east to west, north of the Islands. Seas of 40 to 50 feet or more created in the near-hurricane force blasts passed safely just to the north.

Early on the 23rd a cold front swept through the Islands. Shortly thereafter the surf began to rise in earnest, as north shore boat owners desperately sought to move their craft to shelter in the calmer Waikiki waters. The few foot wind chop at daybreak grew to growling, raving 30-foot monster storm waves by dark.

Daybreak Sunday March 24 revealed unbelievable lines of writhing soup extending to horizon cloudbreaks of at least 40 feet. At times sudden surges of white water overwhelmed beaches, crossed the highway, and damaged homes. Four innocent bystanders near the Pipeline were swept to their deaths. On the island of Hawaii, nearly all buildings of the tiny fishing village of Miloli were obliterated, leaving hardly a trace.

Weakening of the Storm

After the 24th the Pacific storm system drifted slowly north and began to weaken. However the vibrations of wave energy took several days to cross the ocean, creating the excitement of big wave surfing from California to Central America. (Fig. 8-14)

5. Other Sources of Strong Winds

Strong High Centers and Monsoon Winds

So far we have looked at the two major types of wave producers. Sometimes quite strong winds and resulting sizable seas can develop in

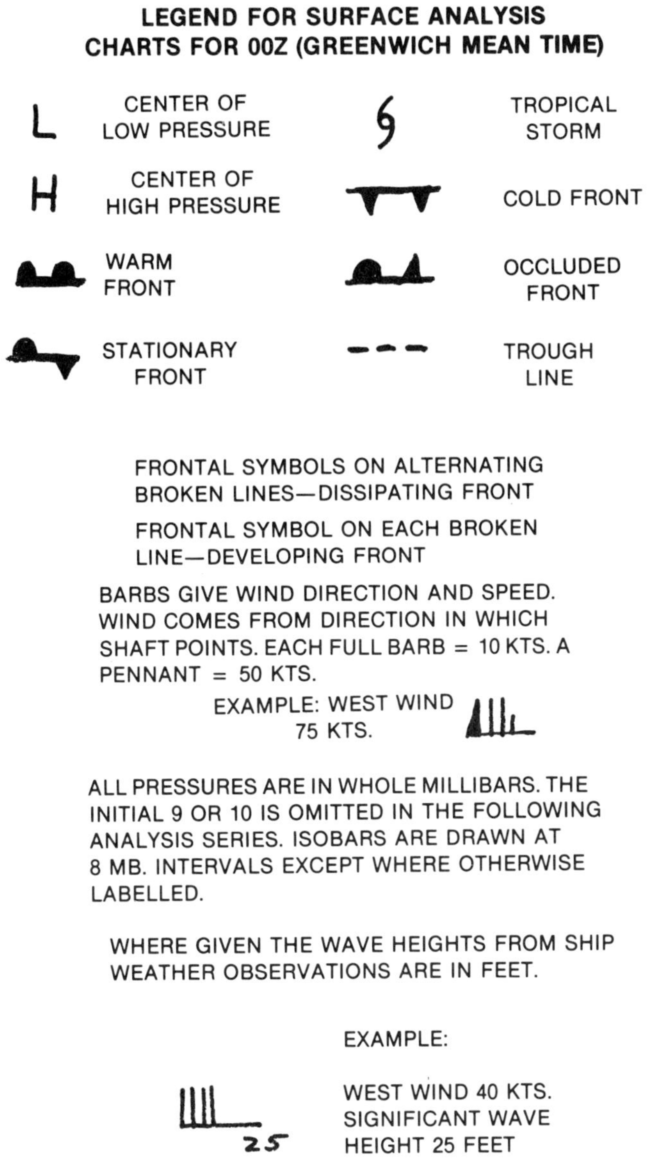

General Weather and Oceanographic Principles

**LEGEND FOR SURFACE ANALYSIS
CHARTS FOR 00Z (GREENWICH MEAN TIME)**

L	CENTER OF LOW PRESSURE	↶	TROPICAL STORM
H	CENTER OF HIGH PRESSURE	▼▼	COLD FRONT
⏶⏶	WARM FRONT	⏶▲	OCCLUDED FRONT
⏶▼	STATIONARY FRONT	− − −	TROUGH LINE

FRONTAL SYMBOLS ON ALTERNATING BROKEN LINES—DISSIPATING FRONT

FRONTAL SYMBOL ON EACH BROKEN LINE—DEVELOPING FRONT

BARBS GIVE WIND DIRECTION AND SPEED. WIND COMES FROM DIRECTION IN WHICH SHAFT POINTS. EACH FULL BARB = 10 KTS. A PENNANT = 50 KTS.

EXAMPLE: WEST WIND 75 KTS.

ALL PRESSURES ARE IN WHOLE MILLIBARS. THE INITIAL 9 OR 10 IS OMITTED IN THE FOLLOWING ANALYSIS SERIES. ISOBARS ARE DRAWN AT 8 MB. INTERVALS EXCEPT WHERE OTHERWISE LABELLED.

WHERE GIVEN THE WAVE HEIGHTS FROM SHIP WEATHER OBSERVATIONS ARE IN FEET.

EXAMPLE:

WEST WIND 40 KTS.
SIGNIFICANT WAVE
HEIGHT 25 FEET

fig. 8

areas of strong pressure gradient along the margins of strong high pressure centers. In the monsoon regions of the Arabian Sea and Bay of Bengal, persistent fresh to strong southwest winds blow during the summer monsoon. In a region off Somalia, each year the summer

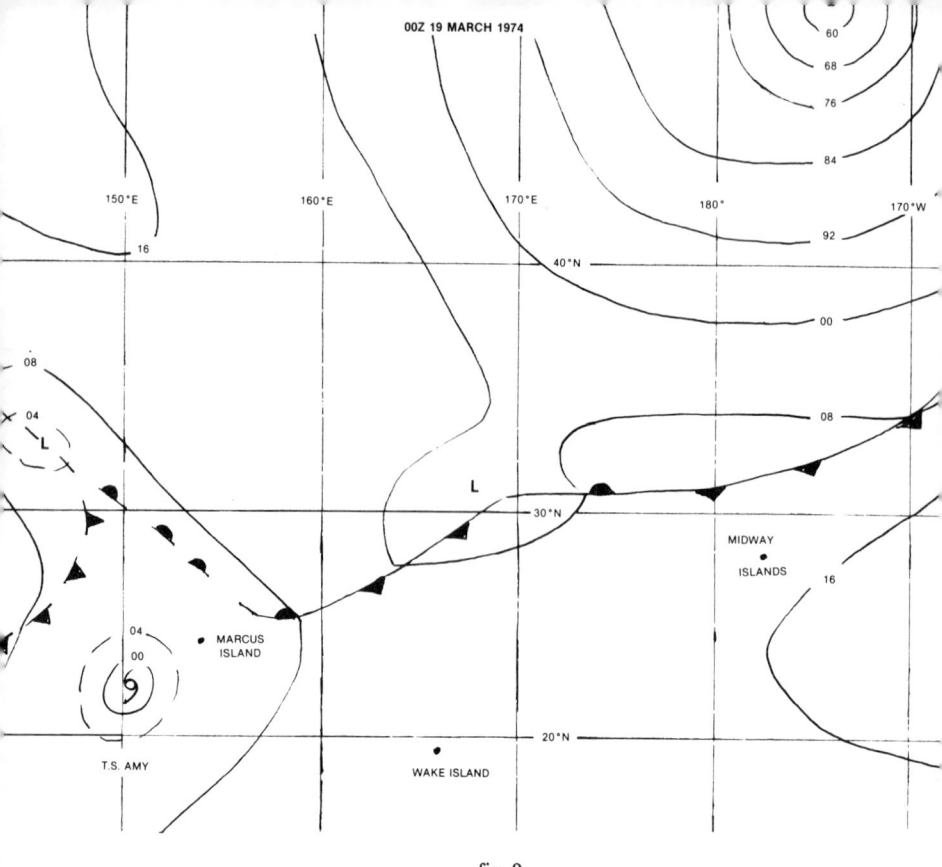

fig. 9

monsoon produces the strongest and directionally most consistent surface winds on earth (mean speeds are approximately 35 knots).

Effect of Local Land and Sea Breezes on Surf

A major factor in determining the wave shape at any locality is the direction and speed of the local wind with respect to the coastline orientation. Many days of potentially good surf have gone to waste because of sudden unwelcome onshores. As we have previously mentioned, the wind blows nearly parallel to isobars with a slight component towards lower pressure. Whenever the pressure gradient (a gradient is the change in a quantity over a given distance) is enough to produce strong winds, the direction of the isobars will give a good indication of the wind direction.

However, when pressure gradients are weak (supporting winds below 10 to 15 knots), the effects of land and sea breeze take over. At night the land cools off quickly and becomes colder than the adjacent ocean. The air over the land becomes heavier than the air offshore and flows towards the sea. This offshore wind is known as a land breeze. By day

General Weather and Oceanographic Principles

the sun heats the land to a temperature much warmer than that of the ocean. Now the ocean air is denser and moves onshore to replace the lighter air over the coast. This is the well known sea breeze. In most areas, on sea breeze days the onshore wind blows from mid morning to late afternoon.

Even if no sea breeze occurs, the daytime heating of land causes a large vertical motion of air. The slower moving surface air is brought into contact with the faster moving air aloft, causing the surface wind speeds to increase. The effects of sea breezes and vertical mixing often cause midday waves to become mushy or marred by choppiness. Usually the glassiest conditions are within 2 to 3 hours of sunrise or in the last hour before sunset.

6. Ocean Temperatures and Currents

Until now, we have considered the factors that produce surf and influence its form. More specific details will be provided in each of the regional sections on waves and weather. Next we will examine the

fig. 10

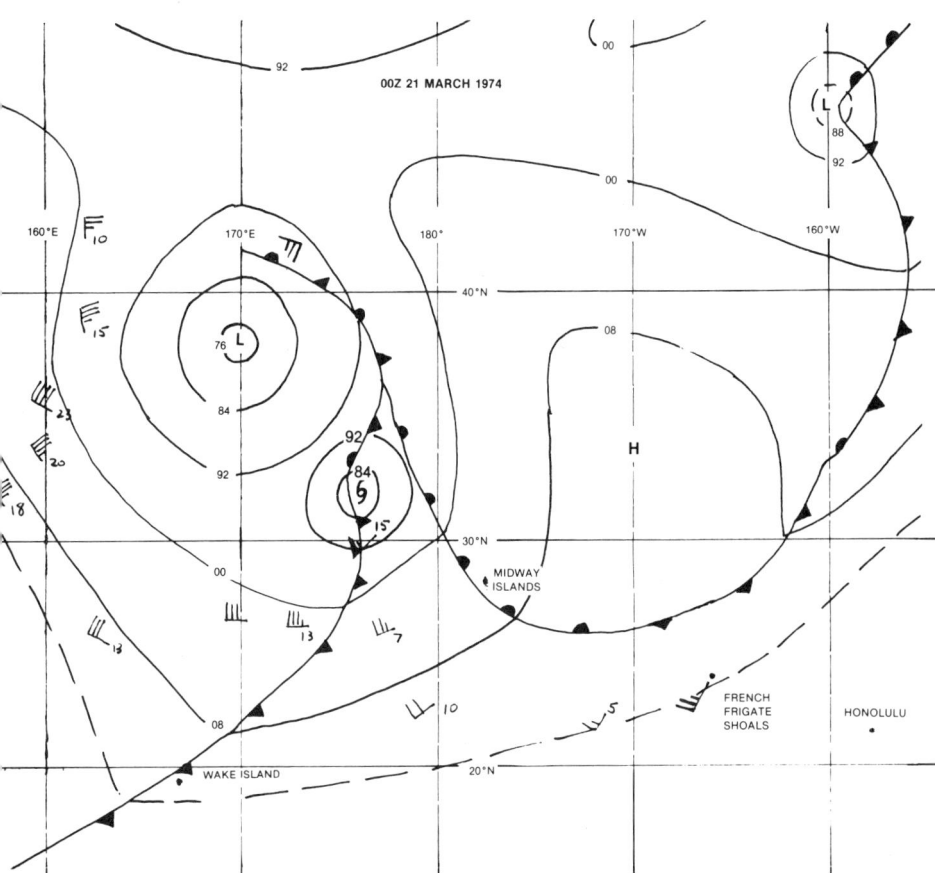

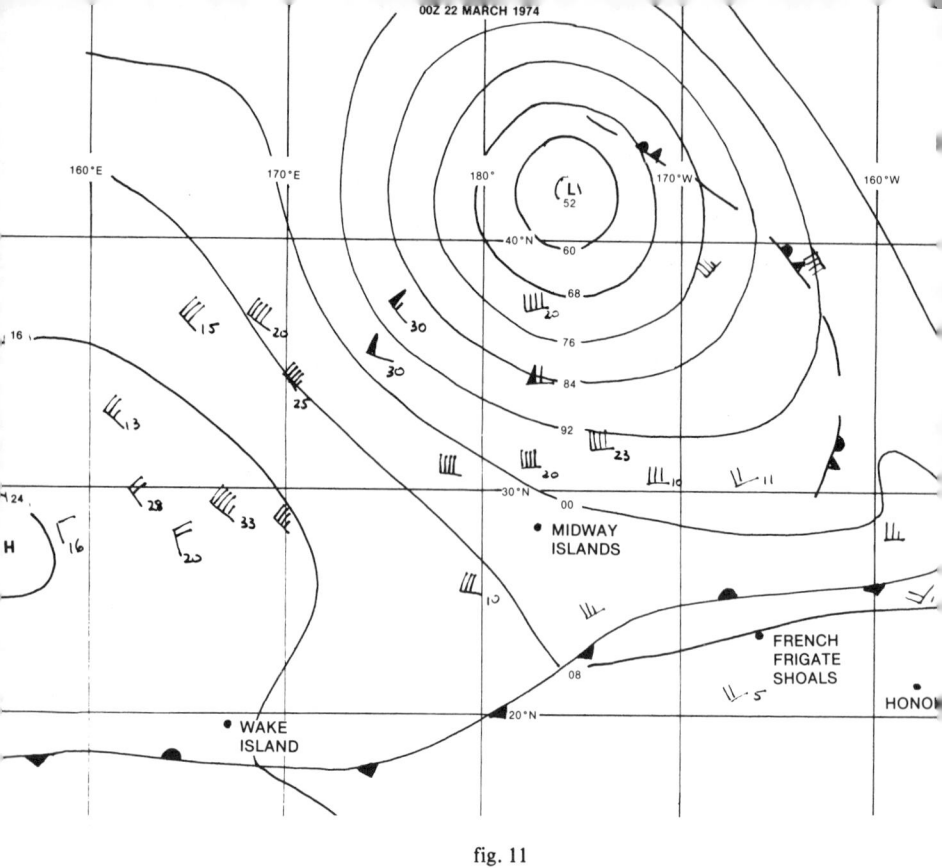

fig. 11

factors that influence the comfort of the surfer. Primarily, these are the water temperature and the climatic conditions of the area of interest.

Global Distribution of Sea Surface Temperature

As would be expected, sea surface temperatures are highest in low latitudes, and with few exceptions decrease poleward. Except for some regions influenced by cold currents the sea surface average is near, or a few degrees, 80° F all year for most of the ocean within 15° to 20° latitude of the equator. At higher latitudes the water is cooler with a definite seasonal cycle of warming and cooling. The annual surface water temperature extremes occur most often in February and August, lagging two months behind the extremes of solar radiation. (Fig. 15, 16)

The greatest seasonal change in water temperature usually takes place near latitudes of 40° to 50°. Poleward of 50° the ocean surface remains relatively cold all year (below 50° F). The strongest seasonal

General Weather and Oceanographic Principles

changes in the Northern Hemisphere occur just off Japan and the area from Cape Hatteras to Nova Scotia on the U.S. East Coast. Refer to the accompanying charts of sea temperature for February and August for greater detail.

Sea Temperature and Individual Comfort

How comfortable an individual feels in any given water temperature depends on factors such as long he has been in the water, how much he moves, how much fat is present on his body, and the temperature to which he is accustomed. Additional meteorological variables including the air temperature, intensity of sunshine, wind speed, and humidity play an important role. Strong wind removes heat directly from the body and helps evaporate water, which causes further cooling.

At ocean temperatures over 80° one can survive immersion indefinitely. Most persons find 80° water very comfortable for surfing or

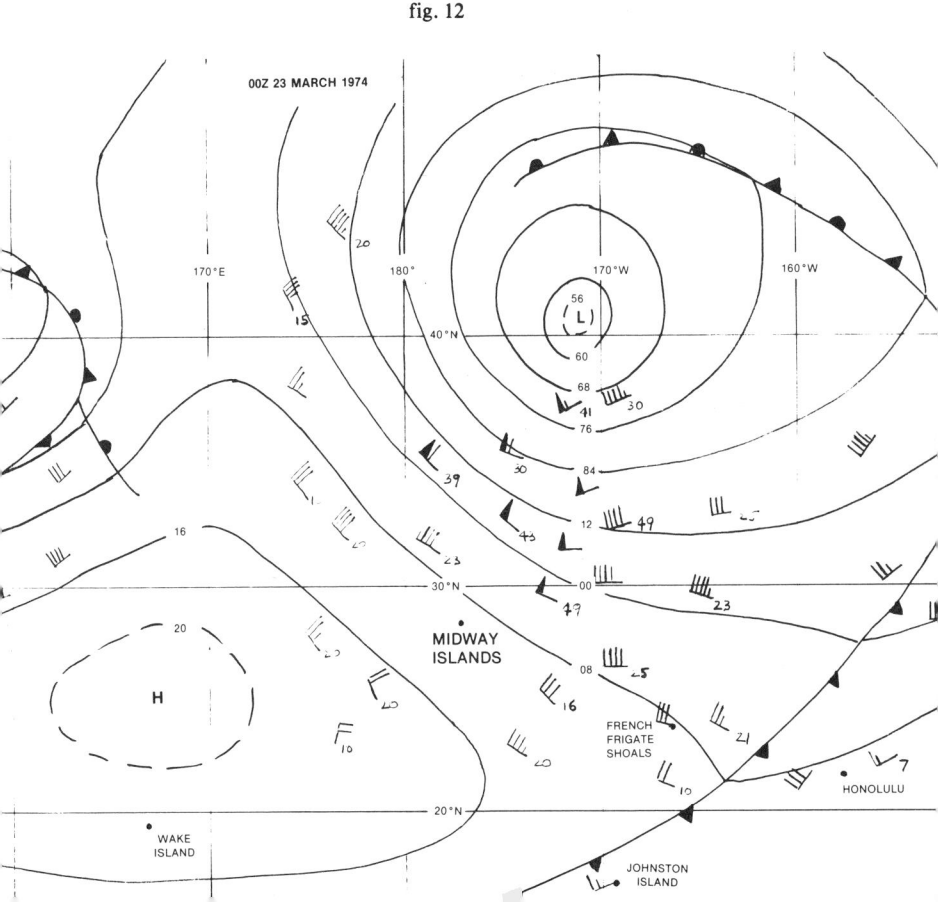

fig. 12

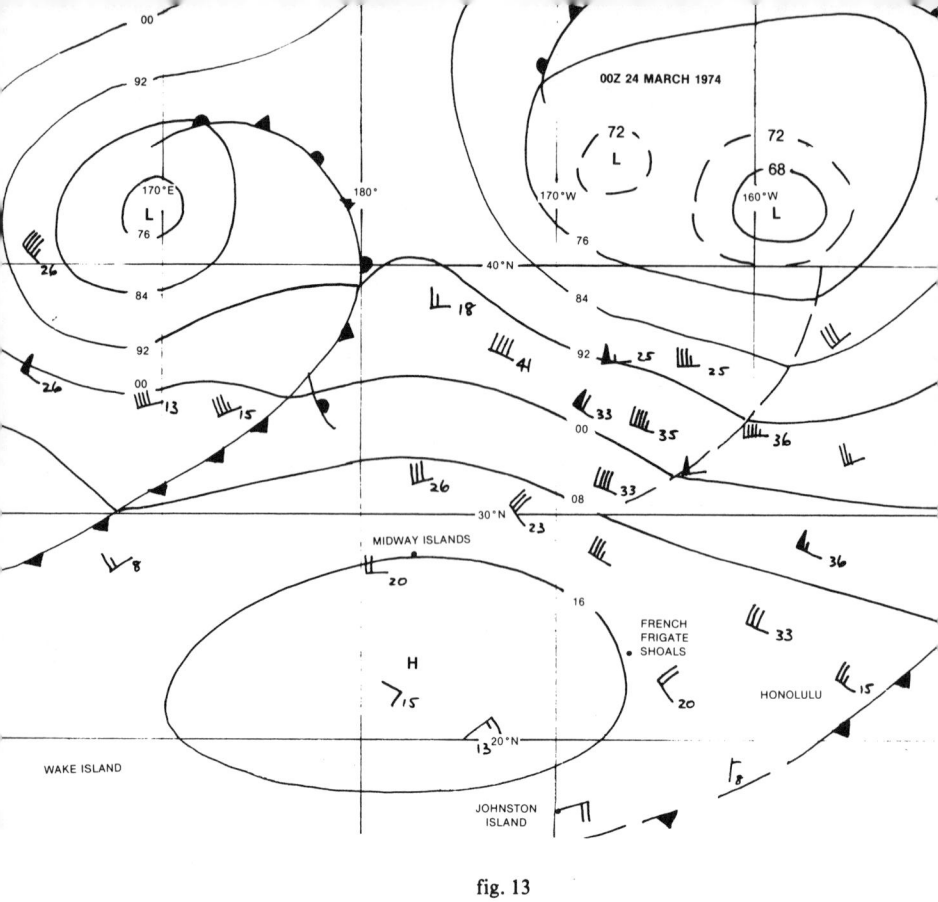

fig. 13

swimming. Water temperatures in the 70s are somewhat more invigorating but still pleasant. Vigorous exercise such as paddling or swimming can still generate enough heat in the unprotected body to keep pace with heat loss to the water and air.

As the water cools into the 60s, most persons will experience a reduction of body temperature with time despite exercise. Eventually unpleasant chilling results, and this may cause the person to leave the water due to the chill before exercise fatigue sets in. Most surfers in 60° waters appreciate the comfort of at least a short-john wet suit.

Water temperatures in the 50s and below cause numbing cold in most unprotected persons after short immersion. Some surfers in year-round cold water regions find water around 55° F to be fairly comfortable with only a short-john. But most prefer full suits in ocean temperatures below the high 50s. And as the ocean chills to the icy 40s

General Weather and Oceanographic Principles

or below, those hardy souls remaining don boots, gloves, and hoods, leaving only their face exposed to the bitter cold air and sea. To ensure maximum comfort and maximize surfing time, one should have wet suit protection appropriate for the intended surfing area. Other reasons for wearing wetsuits include flotation and protection from marine hazards such as coral.

Distribution and Causes of Ocean Currents

As can be seen from the charts, the isotherms of sea surface temperature are not all parallel to the circles of latitude. The reason for this is the transport of water caused by the various ocean currents.

fig. 14

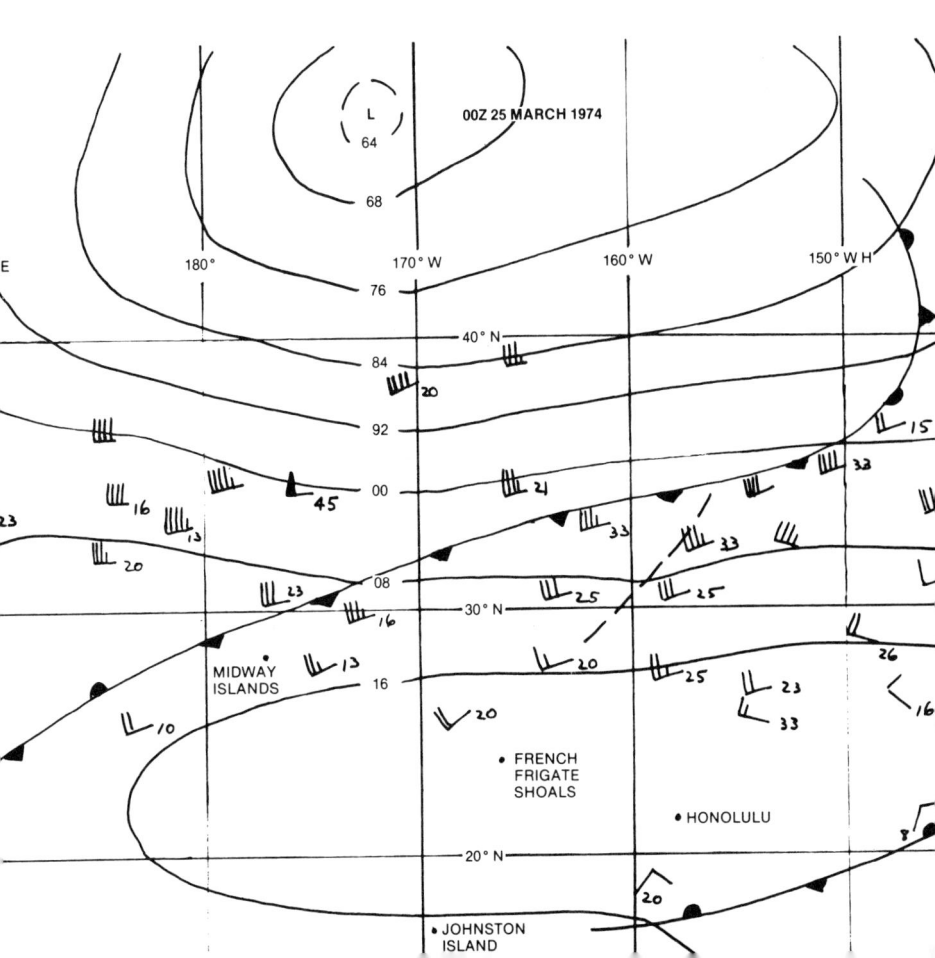

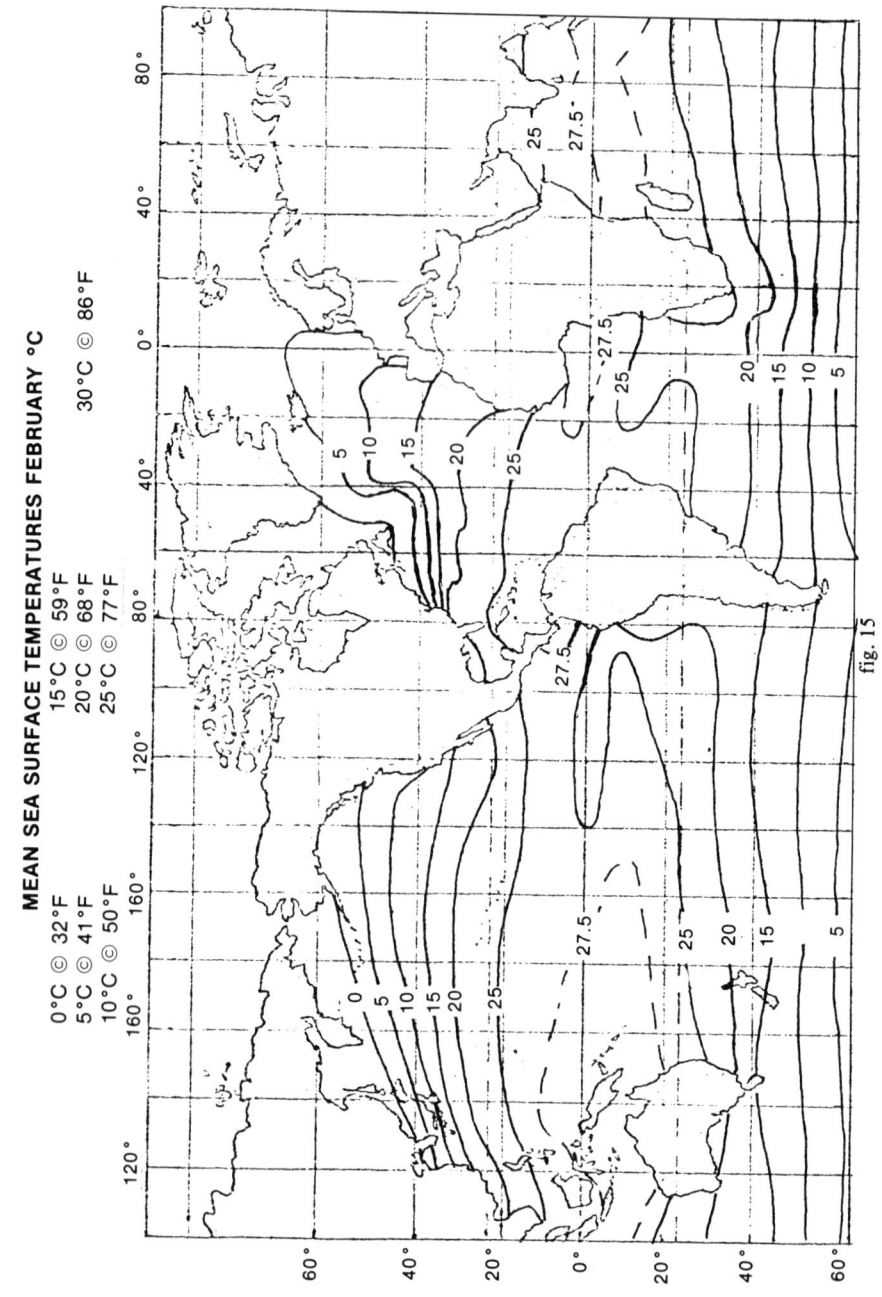
fig. 15

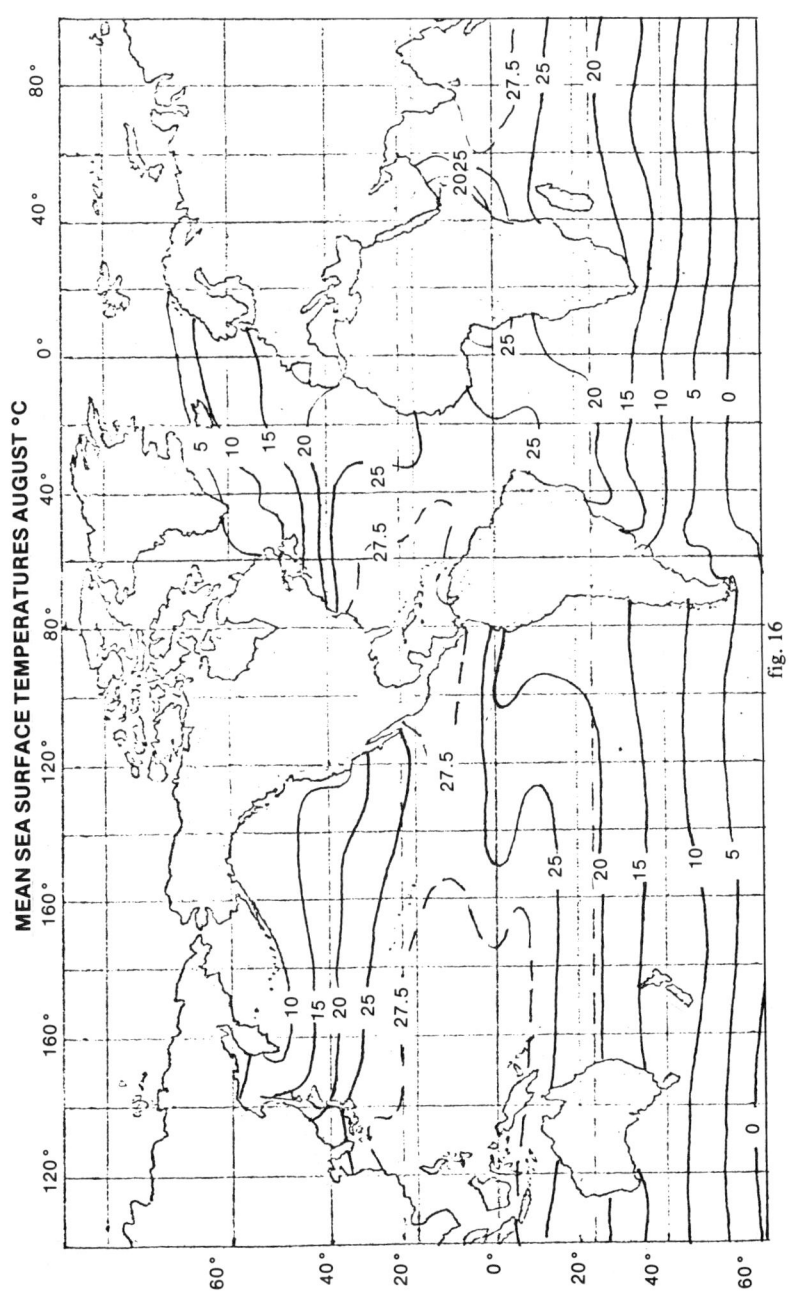
fig. 16

General Weather and Oceanographic Principles

Most ocean currents move in a direction closely related to the mean wind flow over them. With the exception of the Indian monsoon area, large anticyclones dominate the subtropical (near 30° latitude) areas of the oceans. Around these high pressure areas there is a clockwise flow of air and seas in the Northern Hemisphere, and a counterclockwise motion in the Southern Hemisphere.

The anticyclonic flow causes a warm current to flow poleward along the western boundaries of the oceans. As the current approaches 35° to 40° latitude, it turns generally eastward, responding to the westerly winds poleward of the subtropical highs. Near the eastern edges of the ocean the water flows towards the equator, causing a cold current.

In high latitudes and near the equator the current pattern is more complex. Most currents poleward of 45° to 50° tend to curve cyclonically, due to the frequent frontal storms and cyclonic winds of these latitudes.

In the trade wind regions equatorwards of the subtropical highs, the ocean currents move mostly east to west. But, in an area of light winds in the near equatorial trough, there is little wind to move the water along. The water driven westward by the easterly trades piles up along the shores of continents at the western side of the oceans. To balance the increased water level, a return counter-current moves eastward in the light wind area of the trough. (Fig. 17)

One additional effect can cause marked cooling of surface waters. This is upwelling, which is the surfacing of colder deep ocean water. It can cause the sea surface to be 5° to 10° F cooler than otherwise expected. Upwelling often occurs along a coastline where the warmer surface waters are blown out to sea. This causes the deeper water to come up in order to keep a level ocean surface. Upwelling also occurs along the equator in the eastern portions of the Pacific and Atlantic.

7. Climate Over the Oceans

This is a good place to examine the general climate features of the oceans that produce the currents just described. These will assist the reader in understanding the discussions presented in the regional weather and surfing chapters.

Basic Circulation Features

As has already been noted, large anticyclones dominate the subtropical regions of the oceans. A trough within 10° to 15° latitude or less of the equator separates the high pressure cells of the Northern and Southern Hemispheres. Its position varies with the season of the year, although it is usually located near the regions of highest sea surface temperature.

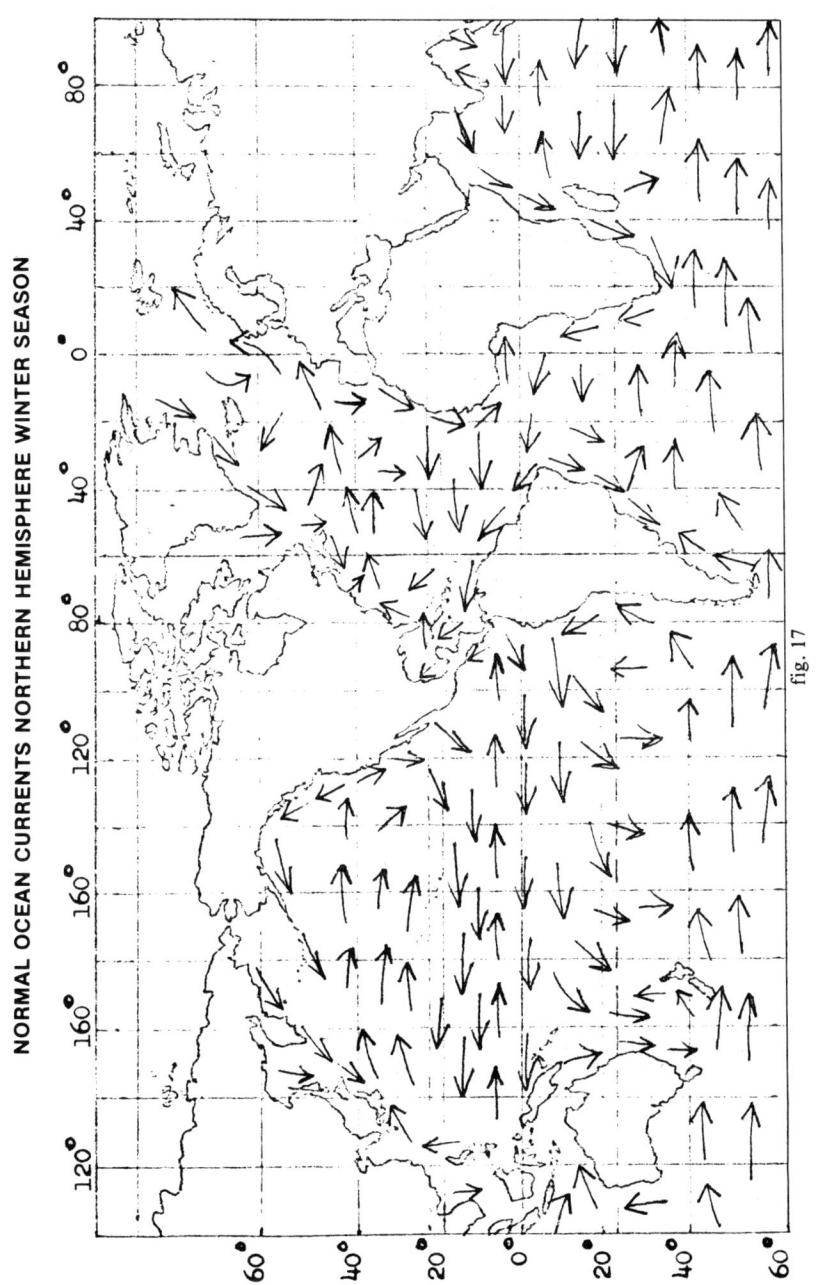
fig. 17

General Weather and Oceanographic Principles

North and west of the subtropical highs, fronts and frequent frontal cyclones travel. The regions between 50° and 60° latitude are particularly stormy, often the producers of very high waves. Polewards of 60°, cold domes of dense air cause high pressure and easterly winds.

The climatology of the monsoon region will be discussed when we consider the Asian and African area.

These features as described are only annual long-term means and show considerable day to day fluctuation in position and intensity. In general, the circulation features are 5° to 10° closer to the poles in summer than in winter. (Fig. 18, 19)

Weather Associated with Circulation Features

Fairly well-defined weather conditions accompany the various circulation features. In the near equatorial trough, the weather is very warm and humid all year with frequent heavy rains and cloudiness. As one moves towards the subtropical highs, fairly steady easterly trade winds and mostly fair weather prevail, with only occasional interruptions by disturbances. The temperature drops only a few degrees from summer to winter.

Near the mean position of the subtropical highs, the winds are light and shift in accordance with oscillations of the high center. The equatorward limits of higher latitude winter storms may cause periods of stormy weather in the cooler season. Annual ranges of mean temperature near 30° to 35° latitude increase westward towards adjacent continents. The winters, particularly near land, have occasional days cold enough to require heavier clothing.

From 35° to 40° polewards to 60° to 65°, lies a major battleground between polar and tropical air masses. Lows and highs move along, each successively influencing the weather. Storms often form along the fronts, especially during the winter. Usually several days of good weather are followed by periods of precipitation. The resulting winds are generally westerly, but shift frequently according to the movement of air pressure systems.

The maximum temperature change from winter to summer occurs between 40° and 50° latitude, and is especially pronounced along the western margins of the North Atlantic and North Pacific. Below freezing temperatures can penetrate several hundred miles offshore during winter polar outbreaks. Summer temperatures over warm currents can be 80° F or more.

Polewards of 50° lies the final destination of most major oceanic frontal storms. Cold, damp, windy conditions with frequent precipitation prevail all year. During the warmer months, mild, moist air from lower latitudes flowing over the cold ocean produces many days of dense sea fog.

MEAN SEA LEVEL PRESSURE JANUARY IN MILLIBARS

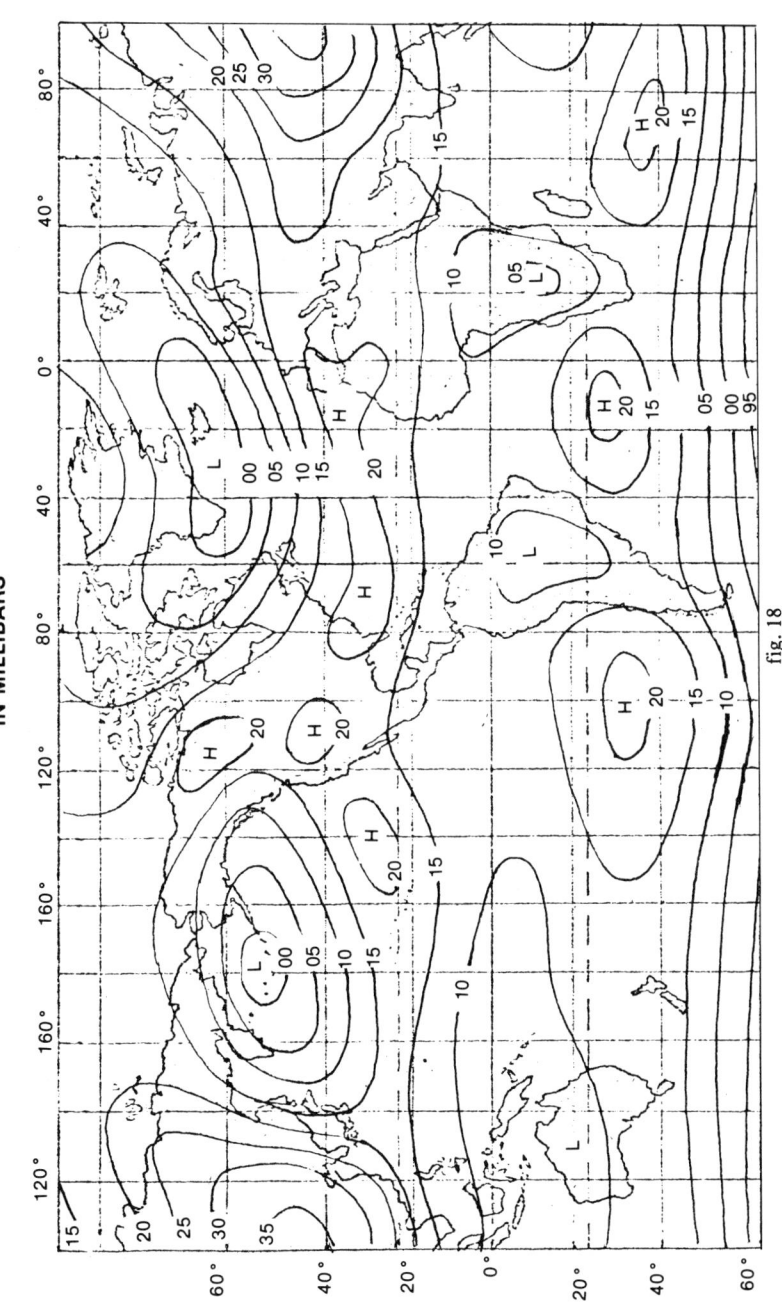

fig. 18

MEAN SEA LEVEL PRESSURE IN JULY IN MILLIBARS

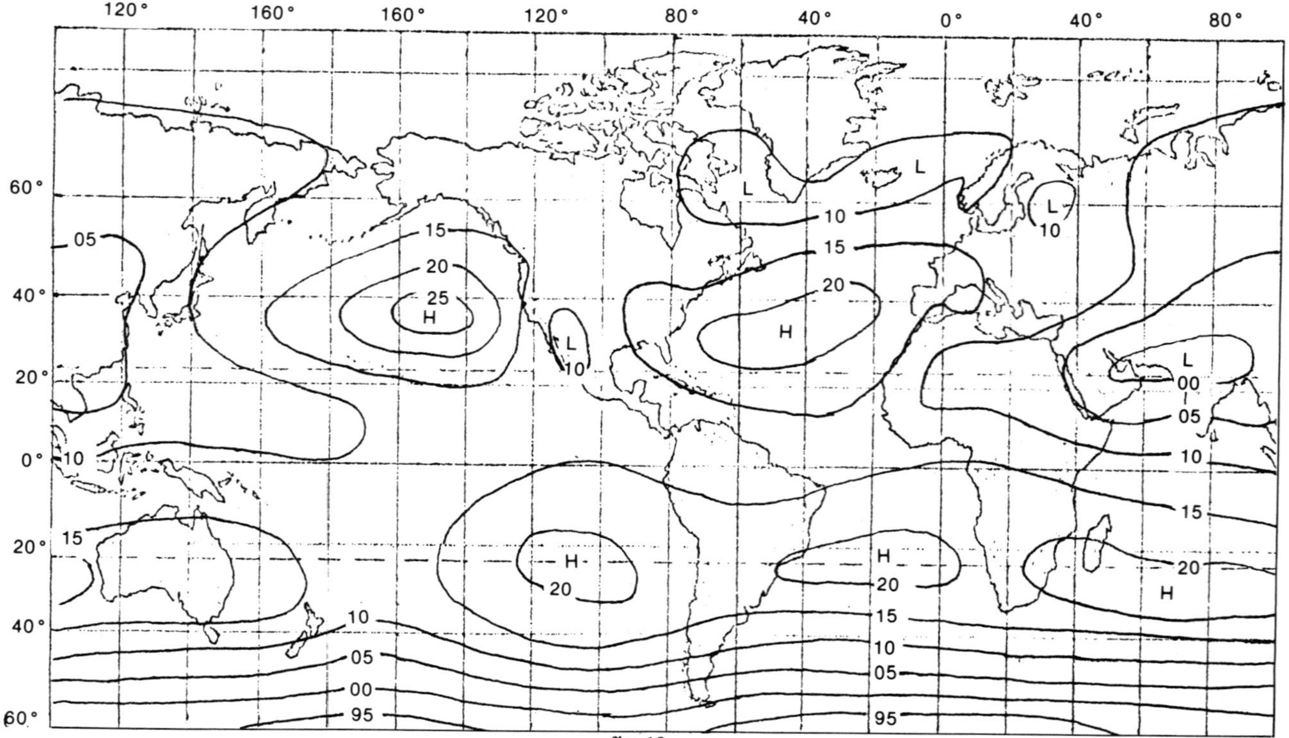

fig. 19

General Weather and Oceanographic Principles

In the polar regions beyond the influence of most storms, precipitation is infrequent. The temperature ranges from a little above freezing in summer to many degrees below in winter. I wonder if anyone has ever surfed under a midnight sun!

Chapter II

Forecasting Weather and Surf

The general principles presented in the preceding chapter provide a basis for estimating the best times to expect surf in a particular locality. But weather and surf conditions change constantly, and a surfer would do well to be his own weather prophet on any extended surf trip.

1. Sources of Weather Information in the U.S.

The amount of information available to make a forecast can range from a keen weather eye to the vast resources of data in a modern weather bureau office. Usually the more information one has, the better the chance to make a successful forecast, especially of surf. But, unless you have a professional meteorologist for a friend, you will have to make do like the rest of us with weather data available to the general public.

Marine Forecasts

Actually, in the U.S., a pretty decent amount of information is available. The National Weather Service prepares forecasts every six hours, or more often if conditions warrant it. Marine forecasts give expected coastal wind, sea, and swell conditions up to 48 hours in advance. Probably the most useful single source of information is the NOAA weather radio which is at 162.40 and 162.55 MHZ on police band throughout the country. It provides a continuous service of specialized forecasts and observations. Radios covering this frequency are fairly inexpensive and well worth the investment. Marine forecasts are also broadcast at 6 hour intervals by the Coast Guard and Marine radio telephone operators around 2.5 MHZ.

TV and Newspapers

Perhaps the next best source is your TV weatherman, especially if he is a professional meteorologist who shows his audience an

Forecasting Weather and Surf

up-to-date weather map with isobars. A number of newspapers publish weather maps, but these are often a half day old by the time the reader sees them. Forecasts on radio and TV news shows are often suspect as some announcers have a tendency to read any forecast on hand regardless of its age.

An unfortunate drawback to nearly all coastal weather reports is the failure to consider weather conditions hundreds or thousands of miles away that might produce surf. Most newspapers have the obnoxious habit of publishing weather maps that extend only a couple of hundred miles offshore, at best. What lurks in mid-ocean remains a vast mystery to 99.9% of the populace.

Tropical Cyclone Warnings

The situation is somewhat better for tropical cyclones. Nearly all sections of the East and Gulf coasts of the U.S. have been visited by a destructive hurricane at one time or another. The public interest in hurricane information is quite high, resulting in frequent broadcasts of tropical cyclone positions, tracks, and intensities. The National Weather Service issues these at six hour intervals, or more often if the storm threatens land.

Along the West coast, beyond the reach of almost all tropical storms, tropical cyclone advisories are seldom given much publicity until heavy surf caused by these storms has already reached the Southern California beaches. However, storm advisories are usually carried on the NOAA weather radio.

An excellent article on Southern California hurricane surf appeared in the July 1974 issue of *Surfer Magazine*.

Local Wind Forecast

In any event, the local weather forecast should provide some indication of the wind conditions expected along the shoreline. This will enable the surfer to figure out when or if the surf may be glassy. An accurate wind forecast is especially important in areas having frequent weather changes.

2. Elementary Forecasting

If one is traveling outside the U.S., the quality and quantity of weather information available will decline, and may become non-existent. A surfer should acquire the knowledge to interpret weather signs from the sea, clouds, and winds along with the aid of a barometer.

The meaning of weather signs varies according to what produces the observed changes in wind and clouds, etc. Roughly speaking, one set of signs applies for tropical latitudes, and another for middle to higher latitude regions. While there is no hard and fast rule, one definition states that the tropics extend to the subtropical ridge line, which averages near 30° latitude. This definition allows for the seasonal shift in circulation patterns. Near the margins of the tropical to mid-latitude boundary one must expect the possibility of either lower or higher latitude weather types. The low latitude type is more likely in summer or autumn, and the high latitude type is more likely in the winter or spring.

Middle and High Latitude Weather Signs

Weather Sequence With a High Pressure Center

In middle and higher latitudes, most weather is dominated by the passage of moving high and low pressure centers. In the Northern Hemisphere an approaching high usually is preceded by northerly winds, a rising barometer, and clearing or clear skies. The air will be colder and drier than normal for the season. (Fig. 20)

As the high center nears, the winds will decrease in speed, becoming westerly if the high passes south of your locality, and easterly if it passes to your north. The rising barometer will level off somewhere above 30.0 inches (or 1016 millibars) in most cases, and fair weather should result.

Along the East coast of the U.S., the easterly onshore flow south of a high center is usually quite moist and may be accompanied by fog and drizzle from the middle Atlantic states northward. Similar easterly flow along the West coast blows offshore from a dry interior. In California the dry warm easterly wind is known as a Santa Ana. This illustrates an important point to be considered in forecasting, which is the source of the air affecting your locality.

As a high center moves away, the wind will become more southerly, and pressures will fall. Warmth and moisture from all levels will move in, frequently with cloudiness and sometimes rain. During the summer months, high pressure often stagnates off the Carolinas, producing a stream of hot, humid air that can last for several days to a week or more in the East and Central states.

Weather Sequence With a Low Pressure Center

Eventually, as a high moves farther away, weather conditions are increasingly influenced by an approaching low or its associated frontal systems. Two sequences of events can occur, depending on whether a

Forecasting Weather and Surf

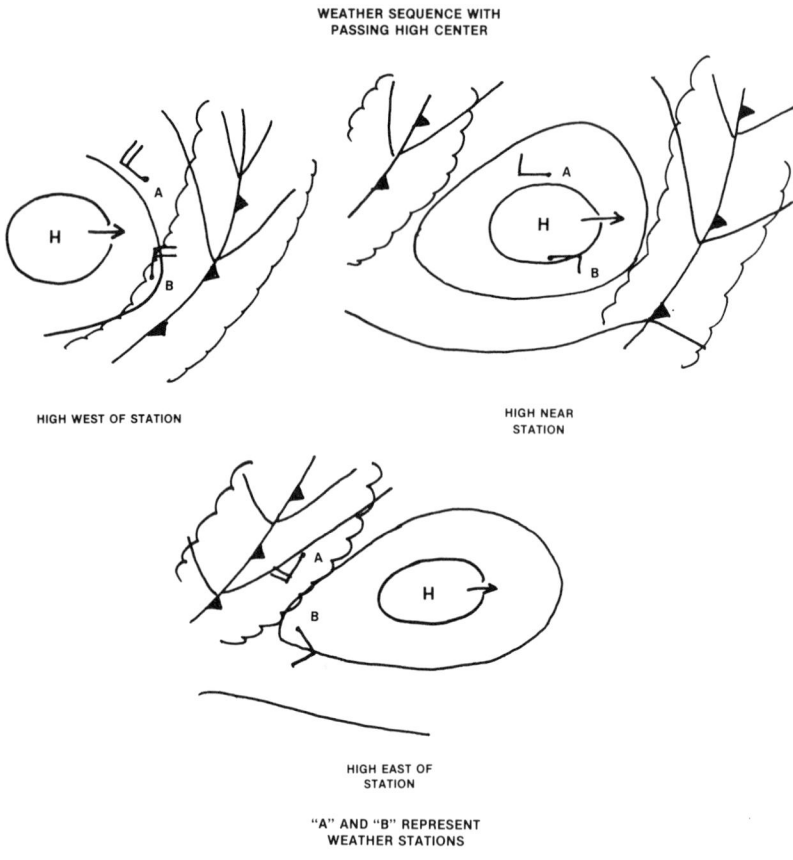

fig. 20

low passes north or south of your location.

If a low passes to the south, the associated winds will begin from the east or northeast and shift through north or northwest as the low moves on by. Steady precipitation should fall. Eventually, after the storm has moved away, drier air will cause the cloud bases to lift. The steady precipitation tapers off, and clearing weather will occur if the low continues to move away. Rising air pressure after a low center passage is good evidence the storm is almost over. (Fig. 21, 22)

The situation is more complicated if the low passes to the north of you. If you are initially in polar air (east of the low center), the first event is a warm front passage. Prior to the front, cloudiness and rain is likely within 200 to 300 miles of the front. Along the West Coast of the U.S., rain may not precede a warm front, as the low level offshore flow

Forecasting Weather and Surf

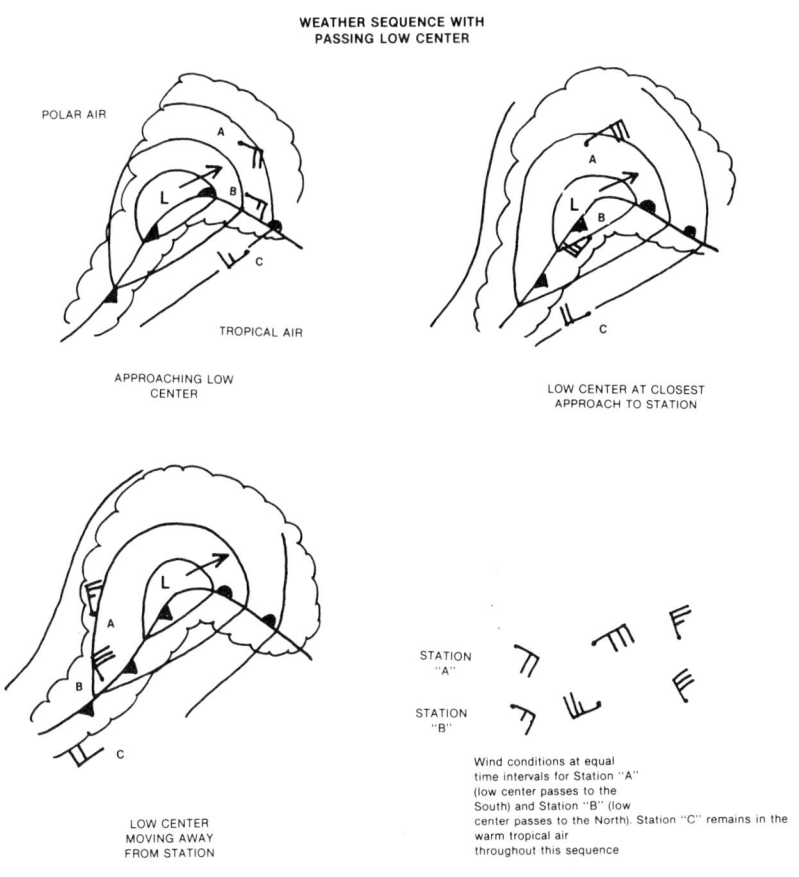

fig. 21

is usually very dry. Pressures fall prior to frontal passage and the surface winds shift from east or southeast ahead of the front to south or southwest after it passes.

Skies often clear partially in the warm sector of the storm after the warm frontal passage. The pressure levels off. The air is from tropical regions, warm and moist. If it crosses a cold ocean surface, dense sea fog is likely. The approaching cold front is next in the sequence. As it nears, the surface wind increases from the south to southwest, and pressures fall again. Lines of showers, thunderstorms, and gusty winds known as squall lines often precede a cold front by several hundred miles.

A strong cold front produces a sharp change in the weather, much more sudden than that produced by a warm front. Showers and

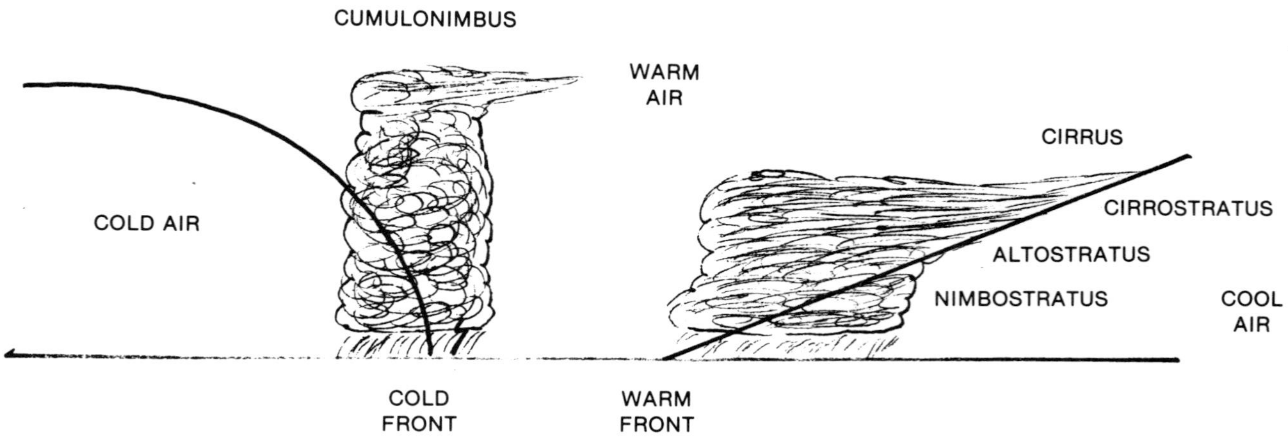

VERTICAL SLICE THROUGH WARM SECTOR
OF FRONTAL LOW CENTER

fig. 22

Forecasting Weather and Surf

thunderstorms often prevail near the cold front. However, some weaker fronts are dry and their passage is marked by no precipitation and only a few clouds. After frontal passage the barometer rises rapidly, the temperature drops sharply, and the wind sets in from a northerly direction. Precipitation usually ends by the time a cold front is 50 to 200 miles beyond your locality.

If a front should stop moving, it is then known as a stationary front. Warm, showery weather prevails near the front on its warm side. In the cold air, layered clouds and steady rain are most likely. Low centers frequently form along cold fronts that have stopped moving.

When an occluded front nears you, the weather ahead of it will suggest an approaching warm front. Layered clouds will thicken and lower, usually followed by steady rain. Winds will increase from the east, southeast, or south. Along the front heavy showers or thunderstorms are likely. Weather after an occluded front resembles conditions following a cold front.

Weather With Cold Upper Lows

Occasionally, bad weather can be produced by what is known as a cold low. This is nothing more than the surface reflection of an intense upper cyclone. No surface fronts are present. Usually these lows are slow-moving and may persist many days. Bands of showers or rain may surround them alternating with relatively cloud free areas. Usually surface winds are light. Upper cold lows produce unstable conditions, meaning that the air cools unusually quickly with height. In such cases, daytime surface heating of land generates heavy cumulus buildups and showers, followed by clearing at night.

Tropical Weather Signs

Consistency of Tropical Weather

In the tropics, weather conditions are dominated by the semi-permanent circulation features that move little day to day. In the trade wind areas, the strength of wind varies with changes in the subtropical high intensity. As a forecast aid, the barometer is generally useless, reading around 29.9 inches, except for the rare approach of a tropical cyclone or strong mid-latitude disturbance. A rapidly falling barometer in the tropics is nearly a certain sign of an imminent severe storm.

Indications of Bad Weather

Tropical weather seldom deviates greatly from what is normal. Usually trade wind skies contain little else but puffy cumulus clouds.

Forecasting Weather and Surf

If a solid cirrus overcast appears and thickens with time, the approach of some disturbance is indicated. Trade wind precipitation consists of light, scattered showers. Steady rains and solid, layered overcast cloud decks are signs of a developing storm. Likewise, if the wind begins blowing from an abnormal direction (easterly winds in the heart of the trade winds area prevail 90% of the time or more), it can be assumed that some sort of storm system is close enough to affect your area.

The discussion in the preceding paragraphs pertains to the Northern Hemisphere. For the Southern Hemisphere, similar events occur, but one must take into account the reversal of wind flow around highs and lows.

The Near Equatorial Trough

In the near equatorial troughs, winds are light and variable. Local winds near land are usually produced by land and sea breezes. All things being equal, more rain over open water falls on days with west winds than with east winds. Day to day fluctuation in rainfall are almost impossible to predict except that very heavy rain is likely if a tropical cyclone is nearby.

Monsoon Regions

The monsoon regions are nearly rain free during the winter monsoon, except for where the air has crossed a considerable expanse of ocean gaining moisture. The summer monsoon is a big rain producer. However, most heavy rains are produced by various types of disturbances within the monsoon. Most of these cause only weak changes in surface pressure or wind, and are therefore hard to detect. Watching the trend in cloud cover types will probably give the best indication of upcoming weather.

Forecasting by Observing the Clouds

High Clouds

If you had to choose one single item on which to base a weather forecast, you should choose to carefully observe the clouds. They give a visual picture of the moisture content and wind direction of the atmosphere from the surface to perhaps 40,000 or 50,000 feet. The clouds can give warning of upcoming weather for time ranges of an hour or two up to a day.

There are 10 major cloud types, each of which carries its own message.

Clouds of the cirrus family or high clouds occur above 20,000 feet.

Forecasting Weather and Surf

The altocumulus and altostratus are mid-level clouds between 6,500 and 20,000 feet. Lower cloud types include cumulus, cumulonimbus, stratus, stratocumulus, and nimbostratus. All of them have bases below 6,500 feet although cumulus and cumulonimbus tops can extend to much higher levels.

Cirrus is a high, feathery looking, thin cloud composed of ice crystals. Good weather is usually associated with cirrus unless it thickens and gives way to lower clouds. Cirrostratus is a solid layer of cirrus through which the sun shines. Often cirrostratus produces solar or lunar halo. A change from cirrus to cirrostratus is a trend suggesting stormy conditions or at least the approaching of a warm front within 24 hours.

Cirrocumulus often forms from a cirrostratus deck as it breaks up due to drier air. These clouds are small, distinct, lumpy masses, often arranged in ripples or like fish scales. The popular name for this cloud is a 'mackerel sky.' Cirrocumulus usually indicate fair weather unless it thickens and lowers.

Middle Clouds

Altocumulus are middle level clouds that appear as a layer of lumpy, ball like masses that tend to merge together. They may appear in patches or align themselves in bands. Altocumulus usually does not precede a general rainstorm, but may be found in the vicinity of scattered showers and thunderstorms. Altostratus is a grayish sheet of middle cloud resembling slate. The sun shines through but dimly. Thickening altostratus usually means steady precipitation is likely within a few hours.

Convective Clouds

Cumulus is a dense, low-based cloud with vertical development. The bases are horizontal, capped by a dome-shaped or turreted top. If the vertical development is not great they are known as "fair-weather" cumulus. Most sunny days, world wide, have at least a few cumulus in the sky. When the vertical development becomes strong, these clouds are known as towering cumulus. Showers are likely to fall from them, especially in tropical weather conditions.

The cumulonimbus cloud marks the extreme of vertical growth. Their towers can penetrate skyward 40,000 to 50,000 feet, or even more. Upper portions of cumulonimbus or "thunderheads" are composed of ice crystals and often form an anvil shaped feature that extends considerable distance downwind from the main mass. Approaching anvils usually mean heavy showers and thunderstorms in the area within an hour or two.

Forecasting Weather and Surf

Note: In case of a thunderstorm a surfer should leave the water and seek suitable shelter. Sea water is a very efficient conductor of the electricity in lightning bolts!

Low Clouds

Stratus is a uniform, gray low cloud whose base is usually a few hundred to 1,500 feet from the surface. Stratus is really nothing more than a fog bank lifted off the ground. The only precipitation from stratus will be a fine drizzle.

Nimbostratus is a low, dark, shapeless layer sometimes having a ragged base. Steady precipitation usually falls from it. These clouds are usually found in the rainy areas preceding frontal cyclones or near organized bad weather areas in the tropics.

Stratocumulus is a cloud intermediate between stratus and cumulus. They are soft, gray, enlongated cloud masses often arranged in long, parallel rolls. Stratocumulus is often the product of decaying cumulus clouds. It usually represents fair weather when present by itself. (Fig. 23)

Other Weather Signs in the Sky

We have suggested a general rule regarding clouds, and now we'll state it more explicitly: steadily lowering cloud bases mean increasing moisture in the atmosphere and increasing chances of stormy weather. Rising cloud bases indicate decreasing moisture and improving weather.

"Rainbow in the morning, sailors take warning. Rainbow at night, sailors delight." This well-known bit of weather lore is a useful forecast aid in mid-latitudes or wherever storm systems move mostly west to east. A rainbow at night appears in the eastern sky, and is caused by the sun shining from the west. The clear skies to the west should soon arrive at your location. Morning rainbows accompany the rising sun. The sun's rays are shining through rain which is already falling just to your west. Shortly this rain can be expected where you are.

Swell Observation as a Forecast Aid

The ocean itself can be a weather prediction tool. The arrival of a swell definitely indicates the existence of a strong wind somewhere. In the tropics during the hurricane season, the onset of a heavy long-period swell is often the first sign of an approaching hurricane. This is especially likely to be true if cirrus and cirrostratus followed by lower

CLOUD TYPES

CIRRUS

CIRROSTRATUS
(Halos Produced Around
Sun and Moon)

CIRROCUMULUS

ALTOSTRATUS
(Sun May Be Dimly
Visible)

ALTOCUMULUS

NIMBOSTRATUS

STRATOCUMULUS

STRATUS

CUMULUS
FAIR WEATHER
TYPE

TOWERING
CUMULUS

CUMULONIMBUS

fig. 23

Forecasting Weather and Surf

clouds invade the sky and thicken, approaching from a direction similar to that of the swell.

Swell direction is best observed in deep water away from shoals, where the waves tend to align themselves parallel to bottom contours. Swell observtions are most useful for forecasting purposes if one knows the probable cause and source location of the swell. With this in mind, one can get an idea of storm movement by noting changes in the swell direction.

3. Basic Swell and Surf Forecasting

The preceding chapters have provided a background on sources of weather data and some rules for forecasting when weather information is limited. An earlier chapter discussed the basics of wave generation. Readers not interested in the technical material of this section should go on to Part II of the text.

Determining Fetch Area

If you should have access to surface weather charts over the ocean, you may wish to do your own surf forecasting. The most difficult thing to do is locate the generating fetch accurately. Try to outline an area which has winds within 5 knots and 15° of an average for a specific time interval. For this purpose surface charts every 6 hours or more are best. This interval is essential if a storm is moving rapidly causing only a short duration blow. But, note that if the fetch is moving directly downwind, the wind will be blowing over decaying waves previously developed upwind. In this case a shorter time is needed to build the seas to their maximum than would be required if the blow had begun over a calm ocean. (The duration time and fetch length required to build seas already present is added to the current wind duration time.)

In any case the fetch can be outlined accurately only if numerous ship observations are available. Frequently ship observations are sparse, as ship masters heed weather advisories and avoid storm areas when possible. Poor estimates of wind speed, duration, and fetch length can easily cause errors of 50 to 100% in swell height prediction. This is especially true if the wind data is derived from isobar spacing rather than actual wind reports from ships.

Computation of Seas in Fetch Area

Assuming you've made a reasonable determination of wind speed, fetch, and duration, you are ready for the next step. Look through the accompanying graphs and determine the approximate height and

Forecasting Weather and Surf

period of seas in the fetch. Remember that for a given wind speed, the wave height will be limited by fetch length or blow duration. The smallest wave height caused by either of the two limiting factors is the one to use.

Computation of Swell Travel Time

Examine the fetch direction to see if it can produce waves reaching your area. When the swell is coming from a long distance (1500 miles or more) be sure to use great circle directions. A great circle is the shortest distance between two points on the surface of a sphere. Be sure to make allowances for protruding coastlines that may block the swells. Next, measure the distance from the forward edge of the fetch to the forecast point. Compute the group velocity of the swell in knots by multiplying the swell period times 1.5. Then divide the speed into the decay distance to get the estimated travel time in hours (or use the graphs). Make allowance for the fact that you've computed the arrival time for the mean period of the swell (the swells containing the most energy). Higher period swell with lower height will herald the arrival of the main swell.

Deep Water Swell and Surf Height at Forecast Point

To figure decayed swell height at the forecast point, use the decay curves (pp. 53, 54). After some experience, you will be able to make rough estimates of the percentage decay by noting the decay distance. If the swell is moving within 20° to 30° directly towards shore, assume the surf height will be 1.5 to 2.0 times the open water swell height, unless experience in your area has shown that some other multiplier is more accurate. Swells making sharp angles with the beach will result in somewhat lower surf but can cause strong long shore currents and rips. Sharply angled swells will cause much higher surf at exposed points than in slightly protected adjacent areas. For the best surf forecasts, repeat the entire process for each successive chart until no more waves are produced that can reach your area.

Instructions For Use of Sea and Swell State Forecast Tables

1. Enter table 1 with wind speed, duration time, and fetch length. Interpolate as needed to obtain required sea states in the fetch.
2. If actual fetch length is less than F Min, use Table 2. This table provides sea state from wind speed and fetch length. Interpolate as needed.
3. To obtain decayed swell height at forecast point, use Table 3. Use significant wave height at end of fetch, decay distance to the

Forecasting Weather and Surf

SIGNIFICANT WAVE HEIGHT AND PERIOD FOR GIVEN WIND SPEEDS AND DURATION TIMES

—TABLE 1—

H ⅓ = Significant wave height in feet
T ⅓ = Significant wave period in seconds
Fmin = Minimum fetch length required to produce above sea conditions

Duration in hours														Sustained wind speed in knots
3		10	15	20	25	30	35	40	45	50	55	60	70	80	100
	H ⅓ FT	1	2	3	5	6	8	10	11	13	15	17	22	26	35
	T ⅓ SEC	3	4	4	5	6	6	7	8	8	9	9	10	11	12
	Fmin N.M.	10	13	16	18	21	23	26	28	30	31	33	36	40	45
6		2	3	5	7	9	11	14	17	20	22	25	31	38	52
		3	5	6	6	7	8	9	10	10	11	12	13	14	16
		23	32	40	48	53	59	65	70	75	80	85	92	100	120
12		2	4	6	9	12	16	20	24	28	32	36	45	54	75
		4	6	7	8	9	10	11	12	13	14	15	16	18	21
		58	77	95	110	130	145	160	175	190	200	210	235	260	300
18		2	5	8	11	15	18	23	28	33	38	44	55	68	*
		5	6	8	9	10	11	13	14	15	15	16	18	20	*
		97	130	160	190	215	240	265	285	305	325	350	390	440	*
24		2	5	8	12	17	21	27	32	37	44	50	63	*	*
		5	7	9	10	11	12	14	15	16	17	18	20	*	*
		140	190	230	270	310	350	380	410	440	470	500	570	*	*
36		3	5	9	14	18	23	30	37	44	50	60	77	*	*
		6	8	10	11	13	14	15	17	18	19	21	23	*	*
		240	310	380	430	510	570	630	690	740	790	830	920	*	*
48		3	5	10	15	20	26	33	41	50	58	68	85	*	*
		6	8	10	12	14	15	17	18	20	22	23	25	*	*
		370	480	570	670	770	860	950	1050	1140	1230	1300	1400	*	*
72		3	6	10	16	22	29	37	46	55	66	76	*	*	*
		6	9	11	13	15	17	19	21	22	24	25	*	*	*
		600	800	980	1100	1300	1450	1600	1750	1880	2000	2200	*	*	*

*Conditions far exceeding naturally occurring values

NOTE: At wind speeds over 45 knots the tops of the waves are blown away as sea spray. A great deal of wave energy is dissipated in foam and turbulence in rapidly developing seas at high wind speeds with short fetches and duration times.

forecast point, and the width of the fetch. This table assumes the fetch is directly pointed at the forecast point. Multiply the resulting decimal by the significant wave height at end of fetch to obtain swell height at forecast point.

4. For swell arrival time, see explanation in text on page 51.

Sea State Near Tropical Cyclones

Determining the sea state around tropical cyclones is often difficult. The main problem is that the wind speeds increase gradually from the outer limits of the storm circulation inward to the eye wall. Seldom is

Forecasting Weather and Surf

SIGNIFICANT WAVE HEIGHT AND PERIOD FOR GIVEN WIND SPEEDS AND FETCH LENGTHS ASSUMING UNLIMITED DURATION TIMES

—TABLE 2—

$H_{1/3}$ = Significant wave height in feet
$T_{1/3}$ = Significant wave period in seconds

Fetch Length N.M.													Sustained wind speed in knots
	10	15	20	25	30	35	40	45	50	55	60	70	80	100
10 ($H_{1/3}$ Ft.)	1	2	3	4	5	6	7	8	8	9	10	13	15	20
($T_{1/3}$ sec)	3	3	4	4	5	5	5	6	6	6	7	7	8	8
25	2	3	4	5	6	8	10	11	12	14	16	18	22	28
	3	4	5	6	6	7	7	7	8	8	8	9	10	11
50	2	4	5	7	8	10	12	14	16	18	20	24	28	37
	4	5	6	7	7	8	8	9	9	10	10	11	12	13
100	2	4	7	9	11	14	16	19	22	25	28	32	38	48
	5	6	7	8	9	9	10	11	11	12	12	13	14	15
250	3	5	8	12	16	19	23	27	31	35	39	45	54	70
	6	7	9	10	11	12	12	13	14	15	15	16	18	20
500	3	5	10	14	18	23	28	34	39	44	50	60	70	*
	6	8	10	11	13	14	15	16	16	17	18	20	21	*
1000	3	6	10	15	21	27	34	41	48	55	62	78	*	*
	7	9	11	13	15	16	17	18	20	21	22	23	*	*
2000	3	6	10	15	22	30	37	46	56	66	75	*	*	*
	7	10	12	15	17	18	20	22	23	24	25	*	*	*

*Conditions far exceeding naturally occurring values

it possible to find an area of nearly constant wind speed and direction that represents a well-defined fetch.

Recently one investigator of Australian tropical cyclones presented a method for predicting wave heights around these storms by consideration of two factors: radius of the eye and maximum wind speed of the storm. Usually storms with large eyes also have a large region of high wind speeds. But information concerning the size of the eye is seldom given in official warnings. Therefore there is little reason to present the relatively complicated method here.

Usually maximum seas in developing tropical storm (35 to 60 knots) are around 15 to 20 feet. As the storm reaches hurricane force some 30-foot seas can be expected. In large and intense hurricanes, seas over 50 feet are possible.

Tropical cyclone seas are nearly always limited in growth by a short fetch or duration. This results in waves that have a short period and are steep. As we have seen before, this type of swell decays more rapidly over a given distance than a long-period swell.

Correction for Storm Motion

The maximum wind speed associated with a tropical cyclone is

DECAY CURVES

—TABLE 3—

Fraction of swell reaching forecast point compared to seas present at downwind edge of fetch

Decay N.M. Distance	Fetch Width N.M.	H ⅓ in feet				Decay Distance N.M.	Fetch Width N.M.	H ⅓ in feet			
		5	10	20	40			5	10	20	40
50	50	.53	.58	.60	.62	2000	50	.08	.09	.11	.12
	100	.61	.67	.69	.71		100	.14	.15	.18	.19
	200	.72	.76	.78	.80		200	.20	.22	.25	.27
	500	.81	.84	.86	.88		500	.26	.28	.31	.33
	1000	.90	.92	.94	.95		1000	.32	.34	.37	.40
100	50	.42	.44	.46	.47	3000	50	.07	.08	.09	.10
	100	.51	.53	.55	.56		100	.12	.13	.14	.16
	200	.60	.62	.64	.65		200	.17	.19	.20	.22
	500	.70	.72	.73	.75		500	.22	.24	.26	.28
	1000	.80	.82	.83	.84		1000	.27	.30	.32	.35
200	50	.32	.34	.36	.38	5000	50	.02	.03	.04	.06
	100	.40	.42	.45	.47		100	.06	.07	.08	.10
	200	.48	.51	.54	.56		200	.10	.12	.13	.15
	500	.58	.60	.63	.65		500	.14	.17	.18	.20
	1000	.68	.70	.72	.74		1000	.18	.21	.23	.25
500	50	.22	.23	.24	.25						
	100	.31	.32	.34	.35						
	200	.40	.42	.44	.46						
	500	.48	.50	.52	.53						
	1000	.55	.57	.59	.60						
1000	50	.13	.14	.16	.17						
	100	.21	.22	.24	.25						
	200	.29	.31	.33	.35						
	500	.36	.38	.40	.42						
	1000	.43	.45	.48	.50						

determined by the pressure gradient just outside the eye. When the storm is moving, maximum winds are increased by VF/2 in the right semicircle (where VF is the forward speed of the storm in knots) and decreased by the same amount in the left semicircle. The opposite is true in the southern hemisphere (see diagram). Also in a moving storm the wind in the dangerous semicircle blows over seas already generated upwind. If the storm is moving at the same speed as the waves it generates, the seas will grow to the maximum permitted under the prevailing conditions of wind speed and fetch length. Wind duration is no longer a limiting factor. As has been noted before, these factors

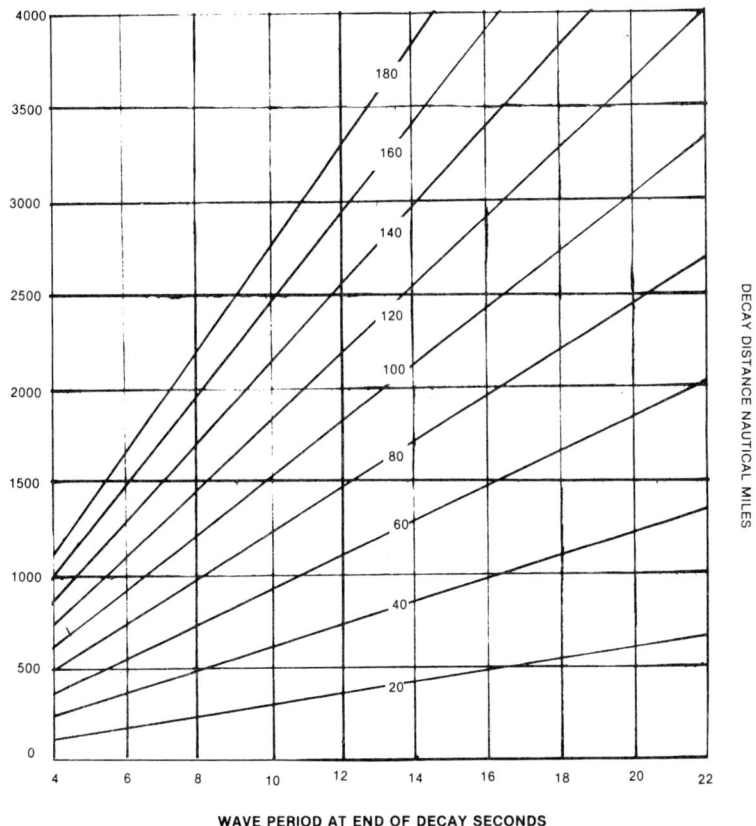

WAVE PERIOD AT END OF DECAY SECONDS

TRAVEL TIME OF SWELL BASED ON
DECAY DISTANCE DIVIDED BY WAVE
GROUP VELOCITY

DIAGONALS REPRESENT LINES OF
CONSTANT TRAVEL TIME IN HOURS

fig. 24

cause the highest seas to be produced in the right semicircle (Northern Hemisphere) in the direction the storm is moving.

If you can make a reasonable estimate of maximum significant wave heights and periods in a tropical cyclone, proceed in the usual manner to compute decayed swell height and surf. Since this estimate is based on the highest seas in the storm, most likely the swell arriving at your locality will be somewhat lower than what these calculations estimate. (Fig. 25, 26)

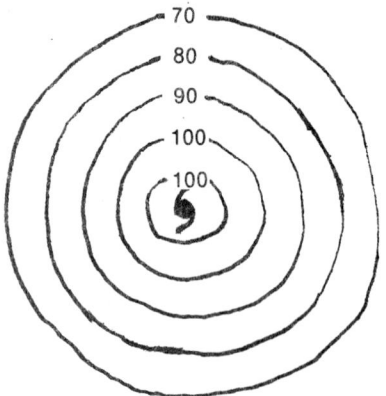

SIMPLIFIED WIND PATTERN AROUND A STATIONARY HURRICANE. WIND SPEEDS ARE IN KNOTS

IN THIS STORM MAXIMUM WINDS ARE 100+ KNOTS. THE WINDS AND SEAS GENERATED AT EQUAL DISTANCES FROM THE CENTER ON ALL SIDES ARE EQUAL.

LET THE SAME STORM MOVE NORTH AT 30 KNOTS IN THE NORTHERN HEMISPHERE. MAXIMUM WINDS TO THE RIGHT OF THE CENTER

© $100 + \frac{30}{2}$
© 115 KTS.

TO LEFT OF THE CENTER MAXIMUM

WINDS © $100 - \frac{30}{2}$
© 85 KTS.

NET EFFECT HIGHER WIND SPEEDS BLOW OVER GREATER FETCH LENGTHS AND FOR LONGER DURATION TO THE RIGHT OF A MOVING CENTER IN THE NORTHERN HEMISPHERE.

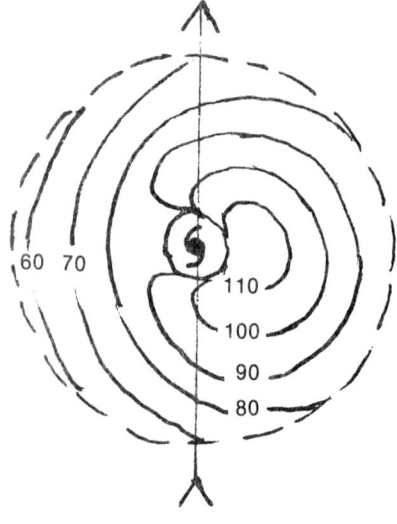

fig. 25

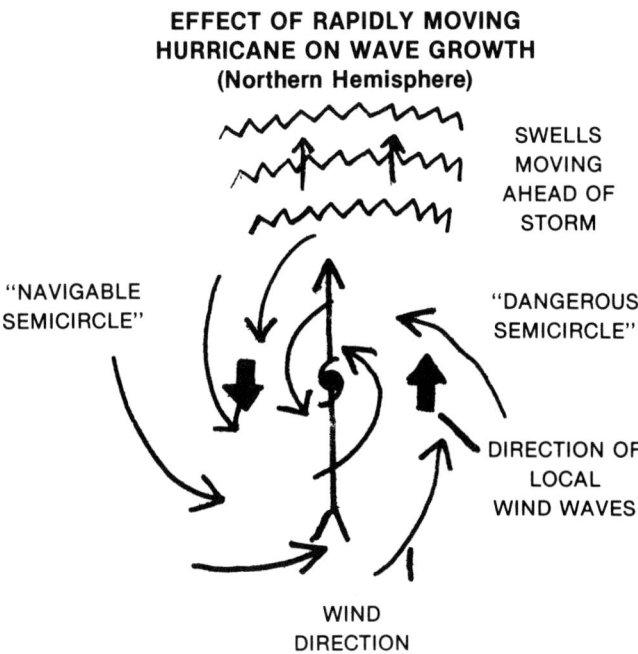

EFFECT OF RAPIDLY MOVING HURRICANE ON WAVE GROWTH
(Northern Hemisphere)

TO RIGHT OF CENTER:
1) Storm is moving at nearly the same speed as the group velocity of the waves. Wave size is limited only by the fetch length of the wind producing them.
2) The direction of the swells preceding the storm and of the hurricane wind waves is nearly the same. The two of them combine to produce higher seas.

TO LEFT OF CENTER:
1) Storm is moving away from the waves it produces causing very limited duration times of hurricane force winds.
2) The wind waves move in a direction opposite to the swell preceding the storm. This results in cross seas where swell and wind waves tend to cancel each other out.

BUT REMEMBER: Conditions can still be very rough left of the center even if not as severe as in the "Dangerous Semicircle."

fig. 26

PART II

Regional Surfing Potential Around the World

Introduction To Regional Sections

If you've followed most of what has been presented so far, you should understand the basic causes of surf around the world. We hope they will help you in planning your next surf trip and make it a success.

Now we are going to take a look at weather and waves for various regions around the world. Our research, while extensive, is by no means complete. If a particular region interests you, careful research should unearth more detail than we can present in this survey. We hope that the preceding chapters have provided the background material to help you in interpreting the needed meteorological and oceanographic data.

In these next sections, we will cover conditions in most known or potential surfing areas of the world. The weather conditions and the resulting surf of the North Atlantic and North Pacific are best known. Portions of the African and Asian monsoon region and most of the Southern Hemisphere (except Australia and some of New Zealand) have not been well explored for surf potential. Neither are the meteorological conditions of the southern waters sufficiently well known to make reasonably accurate surf forecasts.

Organization of the Regional Discussion

A survey such as this of weather and surf conditions world-wide is somewhat difficult to organize so that the material covered is presented in the most logical and efficient way. One method might be to discuss conditions for regions of similar climate. Another might be to analyze what happens month by month. We have chosen to discuss what happens in each major ocean basin and subdivide them as needed to consider regions and seasons having similar conditions.

The North Atlantic and South Atlantic are discussed first. Each is treated separately, as few large swells from one ocean are likely to penetrate far into the other due to the relatively short water distance between South America and Africa. Next we will look at the Indian

Introduction to Regional Sections

Ocean and discuss the peculiarities of the monsoon circulation and its effects on storms and surf. Last is the Pacific Ocean, considered as a single unit, since swells are free to cross the equator from one hemisphere to the other.

Relationship of Climatology to Surfing Conditions

Before discussing the North Atlantic, a few general comments are in order. In many of the following discussions, conclusions concerning expected surf conditions in an area will be derived from the climatology of storms and winds that can affect surf reaching that area. Climatology gives the average weather conditions that can be expected. But, weather patterns sometimes are radically different from the expected state for periods of weeks or even months ahead at a time. As everybody knows, even 24-hour forecasts are sometimes completely wrong. But a surfer with a good knowledge of surf climatology is like a gambler who knows the odds of each possible outcome. In the long run he'll win more often than the guy operating on blind luck!

Differences Between Middle Latitude East Coasts and West Coasts

As each region is discussed, watch the text for recurring patterns that will make it easier to follow. In the middle latitudes of both hemispheres, east and west coasts of continents each have distinctive characteristics. East coasts have prevailing offshore winds and usually quite small and inconsistent surf. Periods of high waves produced by passing storms generally do not last a long time.

West coast winds tend to blow onshore or parallel to the average coastline. Swells generated anywhere in the mid-latitude westerlies may reach the west coast. This results in larger and more consistent surf, at least during the cooler months of the year. Other systematic differences will be pointed out in later discussions.

Chapter I

North Atlantic

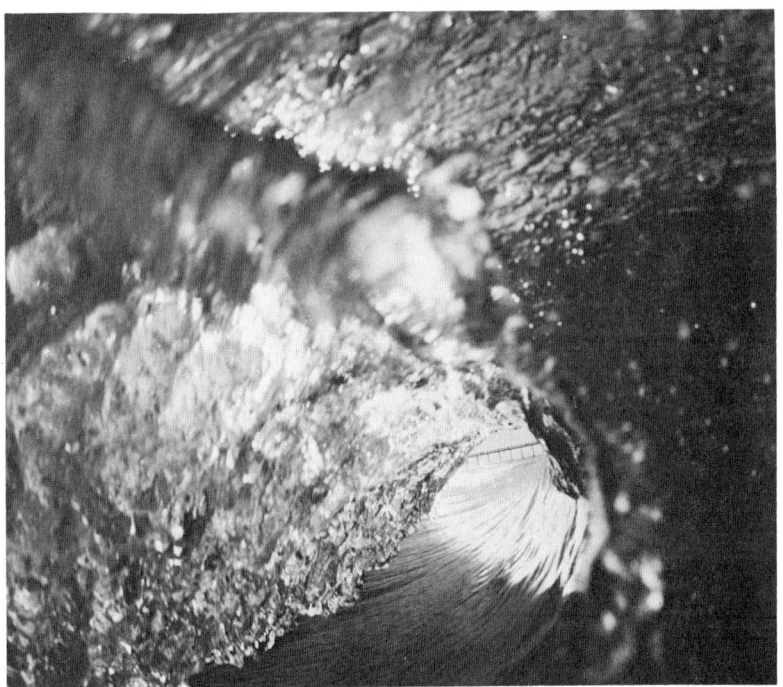

photo: Scott Preiss/SURFING MAGAZINE

1. North Atlantic Wind Circulation

The Atlantic has been important to shipping since the time of Columbus. In fact Christopher Columbus was the best weatherman of his time, giving an unheeded warning of an approaching hurricane at Santo Domingo in 1502. Not surprisingly, the weather and waves of the North Atlantic are the best known of any ocean.

North Atlantic

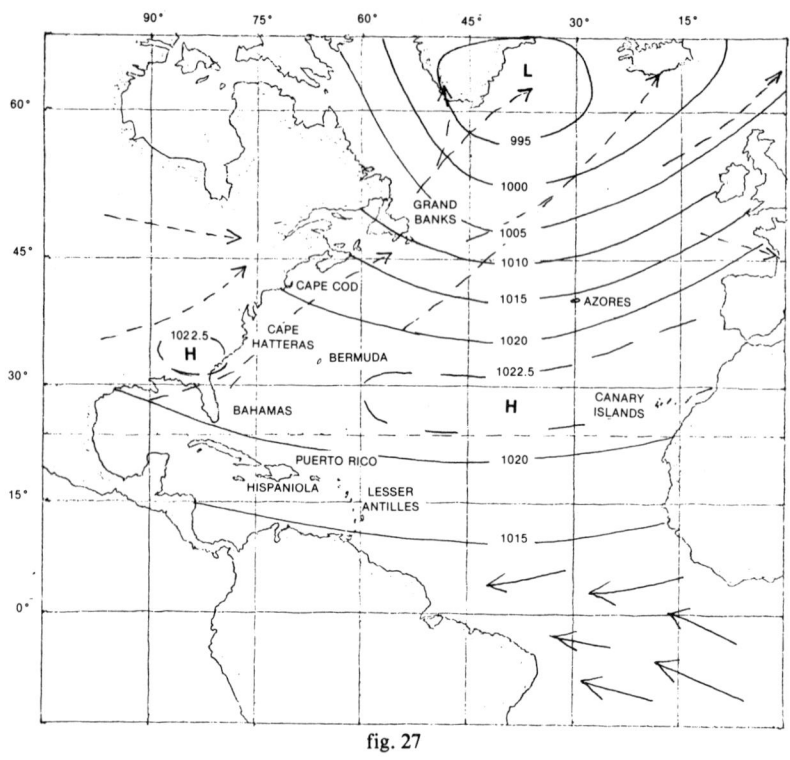

fig. 27

The Bermuda-Azores High

The principal circulation feature of this ocean is the Bermuda-Azores high, a subtropical anticyclone extending from near Bermuda east northeast to near the Portuguese coast. During the summer months the Bermuda high is a steady feature, often ridging westward into the Carolinas or middle Atlantic states and producing long spells of hot, humid weather in the eastern states. During the cooler months, the high center shifts a few hundred miles southward and is often

North Atlantic

NORTH ATLANTIC AUGUST CIRCULATION AND MAJOR SUMMER FRONTAL STORM TRACKS

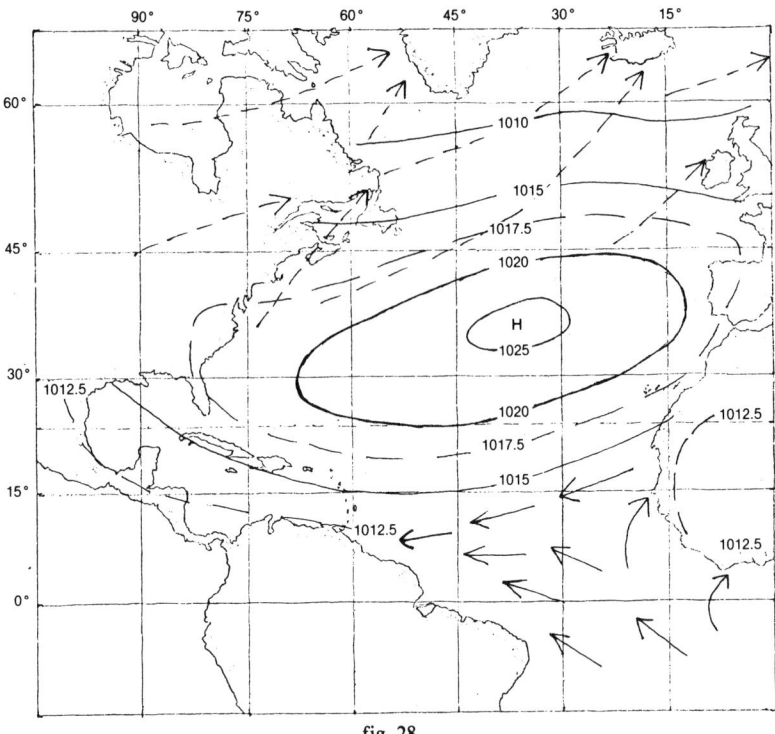

fig. 28

interrupted by mid-latitude storms and fronts. Persistent departures from normal in either position or strength can markedly influence frontal storm and hurricane paths and frequencies, as these systems are steered by the winds blowing around the Bermuda high. North of the high center are the prevailing westerlies and to the south is an east to northeast trade wind region. (Fig. 27, 28)

Frontal Storm Surf

North American East Coast

Most cooler season surf in the Atlantic is caused by frontal storm activity. Frontal storms are most frequent and most intense from

North Atlantic

October and November through March or April. Favored regions for the birth of these frontal storms are waters of the Gulf of Mexico or the Gulf Stream waters from Florida to Cape Hatteras. Every winter at least a few of the rapidly intensifying storms attain hurricane force somewhere between Cape Cod and the Grand Banks area southeast of Newfoundland. During the early stages, the storms are usually rapidly moving, limiting the duration which can build seas. Usually a day or two of large (4 to 8 foot) surf and gusty offshore winds can be expected along most of the Gulf or Atlantic coast regions following one of these storms. Often this type of surf will have a short period with many waves in each set, making it difficult to paddle out.

After a frontal storm reaches maturity and occludes, it generally slows down and expands, allowing high wind to blow over considerable expanses of ocean for a day or more. This raises seas to 30 feet and larger. A stagnating storm offshore may produce up to a week of good-sized waves, particularly along the coastline north of Cape Hatteras which will be exposed to swells generated in the fetch area north and west of the storm center.

Bahamas, Antilles, and N.E. South America

This same situation may produce waves reaching the northerly facing shores of the Bahamas, Hispaniola, Puerto Rico, the Virgin Islands and many of the islands in the Lesser Antilles. Large northerly swells have produced 20-foot surf at Tres Palmas, Puerto Rico, rivaling conditions experienced on the North Shore of Oahu in winter. Northerly swell may also reach portions of South America facing north and exposed to North Atlantic winter storm waves. Despite a probable 1,500 mile or greater decay distance to northeastern South America, a really vigorous North Atlantic gale could raise 10 to 15 foot surf along well-exposed coastal features. Locations in the Antilles or N.E. South America not adversely affected by prevailing trade winds can have excellent surfing conditions whenever a North Atlantic storm swell arrives.

Western Europe and N.W. Africa

Storm activity is also most frequent in the eastern Atlantic during the colder months. Most storms produced in the western Atlantic eventually become part of a semi-permanent low pressure area southeast of Iceland. This Icelandic low is most intense during the winter months. Quite a few storms also develop along frontal systems that have extended into the mid-Atlantic. These will generally move northeastward, and some of the storms will likely reach western Europe. Strong westerly swells produced from higher latitude storms

North Atlantic

will periodically reach the coasts of England, France, Spain and northwest Africa. However, much of the winter surf in Europe will be stormy due to the prevailing onshores of those months. And it will be cold! In more habitable areas of the Azores, Canary Islands, and northwest Africa, the winter months should have the most consistent ridable surf and favorable wind conditions.

Frontal Storms in Warmer Months

Frontal storm activity in the warmer months is usually less intense and confined to higher latitudes. Some summer surf is produced by oceanic storms north of Cape Hatteras to Newfoundland. Some small westerly swells may likewise reach western Europe. During the late spring and early fall conditions are usually in transition between the two main seasons. Storms may become quite intense at higher latitudes, and produce episodes of large surf for a couple of days near or north of 30° to 35° N. But, unless the storm stagnates for an unusual length of time, high surf would not be likely to reach lower latitudes.

2. North Atlantic Hurricanes

Early Season Storms

Fortunately for North Atlantic surfers, during the months of reduced frontal storm activity a second source becomes an important wave maker, the tropical cyclone. The Atlantic hurricane season extends generally from June to November with the greatest frequency of storms in August through October. Most of the early season (up to July) storms originate in the southwest Caribbean in a near equatorial trough extending from the monsoon region west of Central America. Many of these storms move into Central America or the Gulf of Mexico. A few of them cross the southeastern U.S. and regenerate in the western Atlantic. The developing storm can be expected to generate surf while its circulation remains within the Caribbean. Waves will not begin in the Gulf of Mexico until the storm circulation reaches that area. Only a narrow water body, the Yucatan Channel, connects the two water bodies. Several days of large waves and surf are likely in each of these regions whenever a tropical storm or hurricane appears. However the large surf, a welcome change from the infrequent "juice" of the Caribbean and Gulf, may be bumpy due to high winds on the fringes of the storm circulation. (Fig. 29)

North Atlantic

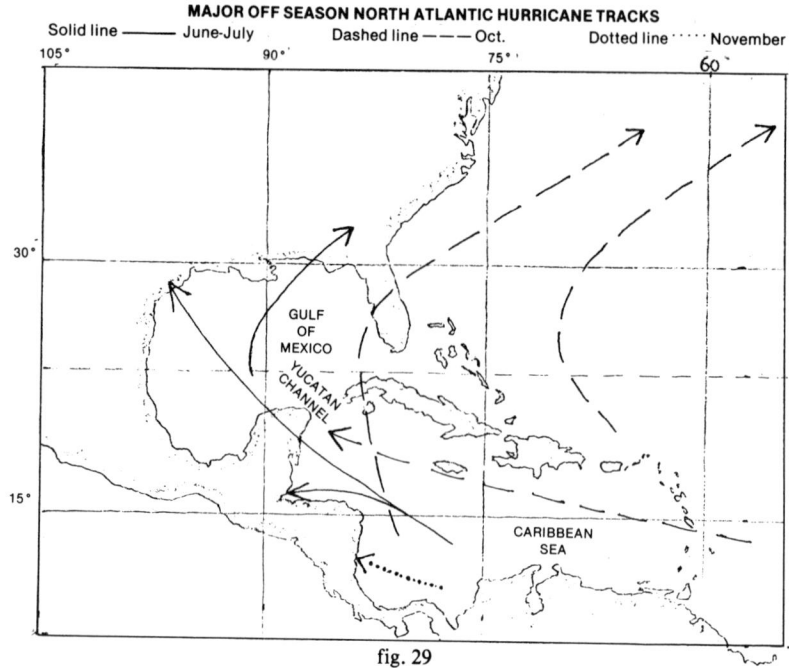

fig. 29

August Tropical Cyclones

August marks the beginning of the main hurricane season. Some storms continue to form in the western Caribbean or Gulf of Mexico, but the most frequent development region shifts to the Atlantic east of the lesser Antilles. Some hurricanes have been traced to circulations originating in the heat trough of the Central North African desert. Many of these mid-season storms traverse long expanses of warm water and become large and intense.

The normally strong Bermuda high forces August hurricanes to move on a generally west to northwest track after they form. As they approach the western Atlantic waters, some will continue westward into the Caribbean, Gulf of Mexico or the southeastern U.S. Others will recurve northward, then northeast, passing between Cape Hatteras and Bermuda. A few hurricanes do not curve eastward and these may cause destructive winds and seas from the Carolinas to New England and the Maritime Provinces of Canada.

Usually, when a storm is preparing to recurve, it becomes nearly stationary for a day or two. High surf produced by a slow-moving western North Atlantic hurricane may last for several days or a week from northern Puerto Rico and the Bahamas and along the coast from Florida to Newfoundland. When the storm follows a normal

North Atlantic

recurving track, winds are likely to be offshore along most of the East coast, producing well formed waves.

September Tropical Cyclones

September is the peak of the Atlantic hurricane season. About one third of all tropical cyclones that develop in the open Atlantic occur during this month. As in August some of the hurricanes become large and intense and follow long tracks over the ocean. These storms may produce up to a week or more of strong swells at suitably exposed beaches anywhere in the Atlantic.

During September there is also an increased probability of tropical cyclone development taking place in the southwest Gulf of Mexico and the western Caribbean Sea. Most of these storms follow courses somewhere between west and north and they are likely to make a land fall after only a few days. Up to three or four days of storm produced surf (or more if the storm moves slowly) may be expected in regions exposed to swells from these storms. (Fig. 30)

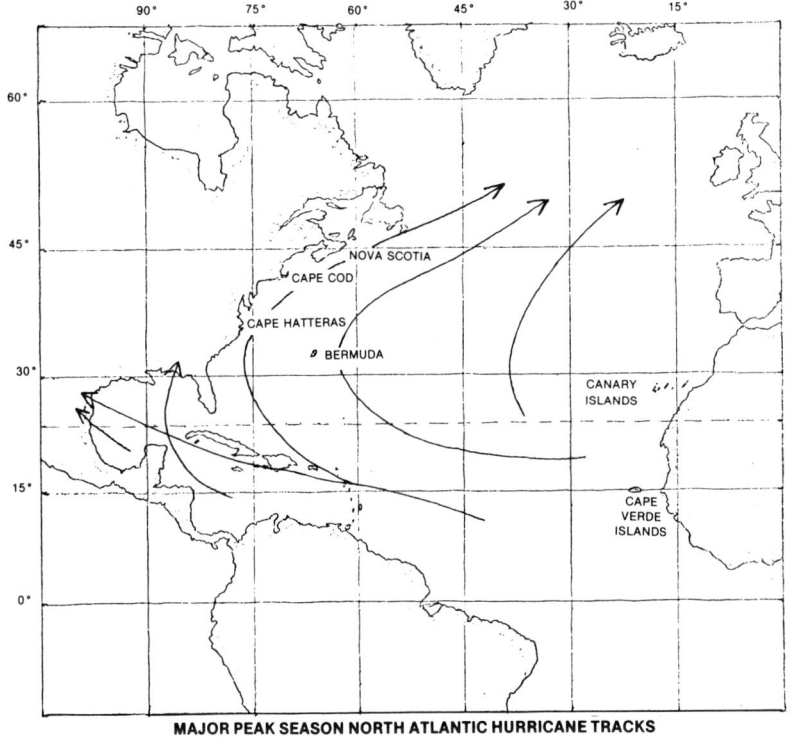

MAJOR PEAK SEASON NORTH ATLANTIC HURRICANE TRACKS
(August-September)

fig. 30

North Atlantic

October Tropical Cyclones

In the latter part of September there is often a slight lull in Atlantic tropical storm formation which is then followed by a secondary frequency maximum in the first three weeks of October. Few October storms develop in the Cape Verde region, but quite a large number develop in the southwest North Atlantic between the lesser Antilles, the Bahamas and Bermuda.

The greatest number of October storms start in the western Caribbean region west of 80° W. Some also form in the southwest Gulf of Mexico. Unlike the earlier season storms, these are more likely to move northward and then recurve to the northeast.

Quite often after recurvature the late season tropical storms become associated with fronts in the higher latitudes of the north Atlantic. This situation will transform an intense small tropical cyclone into a less intense but much larger mid-latitude frontal storm in a day or so. And as is often true, the larger frontal storm may produce higher swells over a larger area than the previous tropical cyclone.

Late Season Storms

Towards late October the Atlantic hurricane season tapers off rapidly. Nearly all tropical cyclone activity is completed by mid to late November. Most very late season storms form in the extreme southwest Caribbean. Some will move onshore quickly, while the remainder will curve to the north or northeast as earlier October storms do. Usually the late season storms will become associated with fronts by the time they reach latitudes of 30° to 35° N.

Variations in Hurricane Frequency

An average of 10 tropical storms occur each year in the Atlantic region. But the annual frequency of storms over the years has varied from one or two up to a total of 21 in 1933. There is also a large year to year variation in the preferred tracks that tropical cyclones follow. But, the odds generally favor periodic episodes of tropical storm-produced surf in the Atlantic, Gulf of Mexico, and Caribbean from August through October. Some times hurricane-produced waves radiate as far as the western European coast.

3. **Other Sources of North Atlantic Surf**

The Bermuda High

The tropical storm and frontal storm seasons explain the causes of most large Atlantic swells, but some smaller and quite ridable surf

North Atlantic

may be produced by wind blowing clockwise around the Bermuda high. Slight southerly swells from this source often cause 1 to 3-foot surf along the east coast from Cape Hatteras northward during the summer months. Offshore winds will usually make this type of surf glassy for a day or two following a cold front.

Caribbean West Winds

During the warmer half of the year a westerly flow is occasionally found in the western Caribbean. If strong enough, these westerlies may produce good surf on the western shores of the islands in the eastern Caribbean. There is also a narrow belt of southwesterly monsoon flow in the extreme eastern Atlantic south of the near equatorial trough in summer and early fall. Parts of Africa north of $5°$ to $10°$ N might receive some surf from this source.

4. Other Important Factors in Evaluating Surf Potential

Local Wind Conditions

Most basic causes of North Atlantic surf have been described in the previous sections. For each place the surfer will need to take into account the effects produced by the local wind and bottom conditions. Regions in higher latitudes will experience frequent wind shifts due to passing high and low pressure areas. In the Atlantic trade wind belt, easterly exposed beaches will have poor, mushy, wind-blown conditions much of the time. And finally, on sunny, light-wind days when the ocean is colder than the land, there is a good chance of a sea breeze producing onshore winds. On probable sea breeze days the best time to surf (tides permitting) will be the first couple of hours in the morning and again near sunset.

Bottom Conditions

The Continental Shelf

Bottom conditions play an important role in the Atlantic in determining shape and speed of the waves. Along much of the U.S. East and Gulf coasts the continental shelf extends 100 to 200 miles seaward. The water depth is shallow enough so that longer period waves feel the bottom and slow down.

But southeast of Cape Hatteras, the continental shelf only extends just past Diamond Shoals, 20 to 30 miles offshore. The combination of a pronounced point and deep offshore waters causes Hatteras to have some of the strongest and most consistent surf along the U.S. East coast.

North Atlantic

Shallow Water Near Shore

Near shore in many places the water depth drops off very gradually. Long lines of parallel sandbar formations occur, changing constantly in shape with the flow of tides, waves, and currents. The combination of an extended continental shelf and gently sloping inshore bottoms are apt to produce spilling, relatively slow waves. But, there are places with rock bottoms that can break with great force. The most rock-bound region in the western Atlantic extends from Maine to Newfoundland.

Chapter II

South Atlantic

The next ocean area we will discuss is the surf potential of the South Atlantic ocean. Compared with weather and oceanographic information for the North Atlantic, information about the South Atlantic is sparse and somewhat unreliable. Only in recent years have satellite pictures provided data on the frequency of storm activity. But the basic principles applying to all major ocean regions will give a general picture of storm and swell conditions likely in possible surfing areas bordering this ocean.

1. South Atlantic Wind Circulation

The primary circulation feature of this ocean is the South Atlantic high, which is centered on the mean 5° to 10° W and 25° to 30° S moving only slightly poleward in summer. Very persistent SE trade winds and a lack of major storms prevail north of this feature.

Lack of Tropical Cyclones

The near equatorial trough of the Atlantic generally lies in the Northern Hemisphere except from February to April when it is occasionally observed between 0° and 5° S. Because of this tropical cyclones are almost unknown in the South Atlantic, and there is no low latitude source of large surf.

2. "Roaring Forties" Frontal Storms

Near and south of the mean positions of the South Atlantic high, fronts and frontal storms are observed, especially in the winter months. Cyclones are most frequent in the latitudes of 40° to 50° S. This is the so-called "roaring forties" belt in which high speed westerly winds occur frequently all year. Quite a few storms initially form in the southwest Atlantic east of Argentina, and develop as they move eastward. The southward flow of cold southwesterly wind in

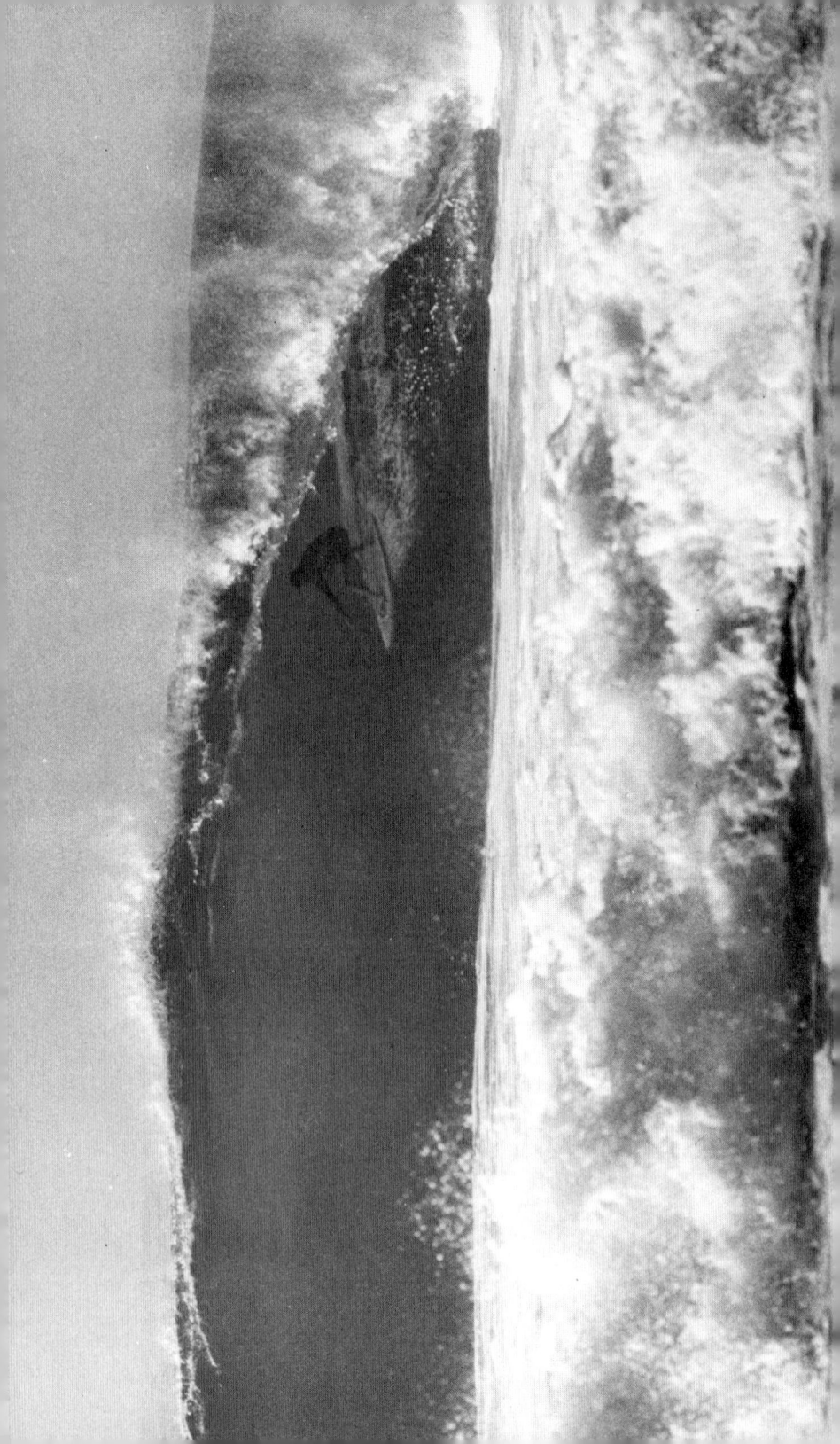

South Atlantic

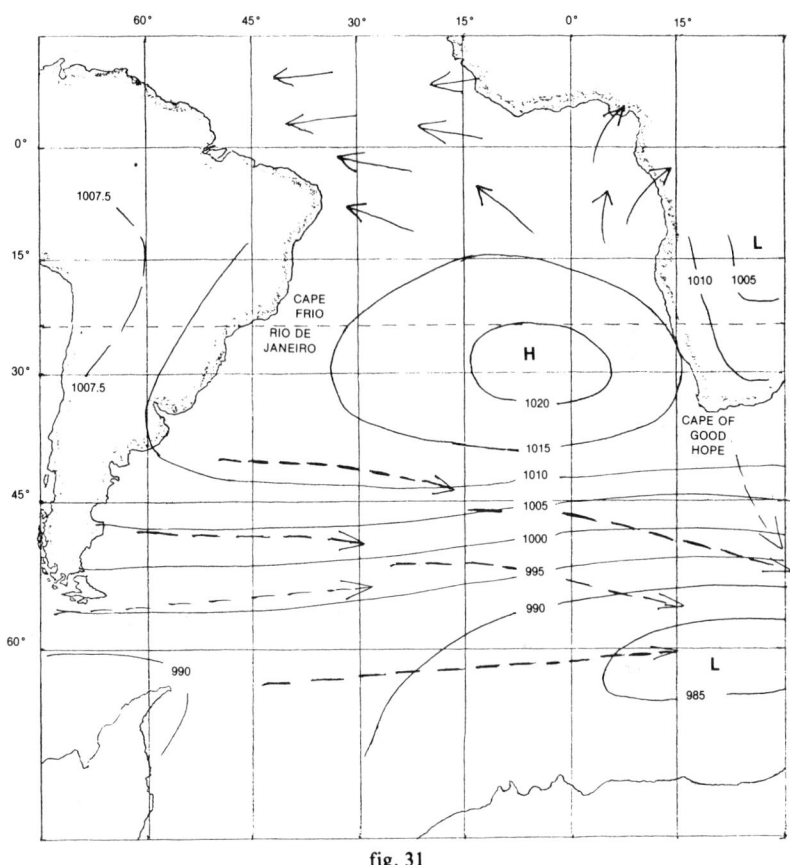

fig. 31

winter sometimes extends to latitude 20° S or even 10° S. Likewise a considerable amount of cold front activity is likely in winter (June through August) in South Africa. (Fig. 31, 32)

3. Geographical Effects on South Atlantic Surf

Low Latitude Eastern South America

After considering the geography of the land masses surrounding the South Atlantic it is possible to derive the following conclusions from

South Atlantic

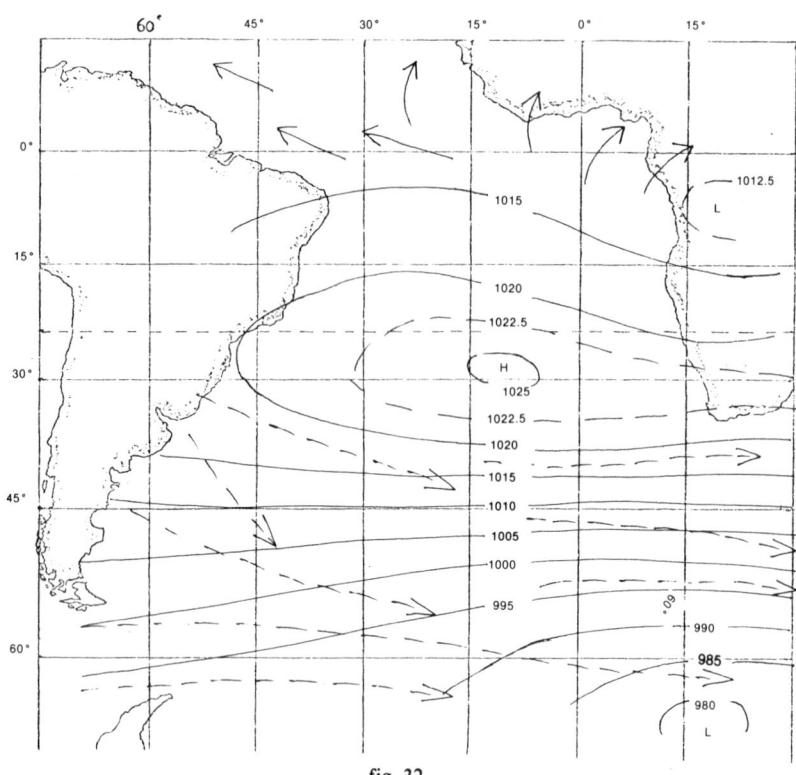

fig. 32

climatology: first, little surf from a South Atlantic source is likely along the South American coast north of about 5° S, as this coastline faces toward the north-northeast. (But winter surf from North Atlantic storms is likely at times—see preceding chapter.) Not much well-formed surf is likely from latitudes 5° S to near Cape Frio, Brazil, as this region lies in an area of persistent southeast trades and a lack of storms. Some small southerly or southwesterly swells may cause surf in winter when strong surges of cold air reach low latitudes in the western South Atlantic. The jungle-like conditions of parts of Brazil, the dangerous marine life (sharks, etc.), and probable lack of consistent surf do not seem to favor these regions for surf trips.

Middle Latitude Eastern South America

Near and south of the Tropic of Capricorn (23½ ° S) the eastern South American coast line becomes irregular with a number of points

South Atlantic

exposed enough to receive the effects of southerly swells. Well-formed (small to medium sized) surf is quite common in winter at Rio de Janeiro and vicinity. The sub-tropic climate and warm water conditions make conditions appear to be frequently agreeable along the eastern Brazilian coast from near 22° S to about 30° S in the cooler months of the year (May through October).

In higher latitudes the size of the surf is likely to be larger but the frequency of wind-blown, choppy conditions is also greater. And polewards of 35° S, the water will be icy cold (below 50° F) during the winter and only a few degrees higher in summer. Some good cold water surfing will happen when swells arrive from the southern portions of South Atlantic storms.

Southwest Africa

The African side of the South Atlantic appears to hold good potential for consistent large surf, especially south of about 25° S. Storms are frequent in the Southern ocean all year and many strong southwest swells approach the southern tip of Africa. Due to the Benguela current, the water is cold along this coast as close to the equator as 15° to 20° S, and fog is likely. The largest waves should be expected in May through October, but onshore winds from cold fronts are likely to make quite a few days choppy. Somewhat smaller waves but better wind conditions should occur the rest of the year.

Equatorial West Africa

Close to the equator the surf will be smaller and not as consistent as in southwest Africa. But, some good tropical water surfing is likely along the Gulf of Guinea (actually in the North Atlantic) and the other low latitude regions of South Africa exposed to southwest swells of the "roaring forties." However, southwest onshore winds will be frequent in the Gulf of Guinea from May or June until October. The many points and reefs and long expanse of unpopulated coastline make most of the South Atlantic coast of South Africa a good bet for surf exploration.

Chapter III

Indian Ocean

1. The Monsoon

The next major oceanic area is the Indian Ocean. Nearly all places of potential interest to surfers in this area have fairly warm water all seasons. Although it is one of the smaller oceans, the weather features are quite complex due to the dominance of the Asian monsoon circulation. In simplest terms, the monsoon develops because of large-scale differences in the temperature between the Asian continent and the adjoining oceans.

Northern Hemisphere Summer

During the Northern Hemisphere summer the air over Asia is strongly heated by the sun. The heated air rises and cooler air from the ocean moves inland to replace it. The subtropical high of the Indian Ocean lies near 30° S in the months of June through September. Steady southeast trades prevail from there to the equator. In the Northern Hemisphere strong southwest to west winds blow into the low pressure over Asia.

Northern Hemisphere Winter

In the Northern Hemisphere winter season, Asia is cold and northeast winds blow over the Arabian Sea and Bay of Bengal. The winds tend to become northwest within a few degrees south of the equator east of 50° E blowing towards low pressure over Northwest Australia. From 15° to 20° S to the latitudes of 30° to 35° S the wind continues to be southeast trades.

Transition Seasons

The months of April and May plus October and November represent a transition between the two extremes of the Asian monsoon

Indian Ocean

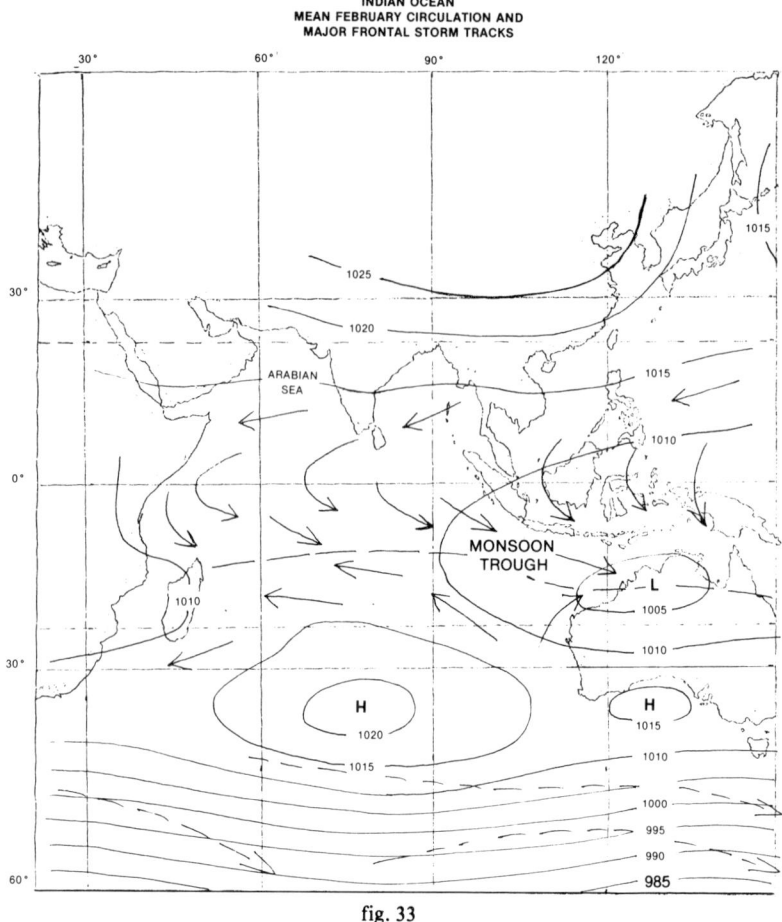

fig. 33

circulation. Winds in the Arabian Sea and Bay of Bengal are light and variable and tropical cyclones may form. Once in a while a pair of simultaneous tropical cyclones may appear near the equator in the Indian Ocean at the same longitude in each hemisphere. (Fig. 33, 34)

2. Tropical Cyclones

Northern Hemisphere

About eight tropical cyclones form per year in the Bay of Bengal and Arabian Sea, mostly during the transition period. A storm or two

Indian Ocean

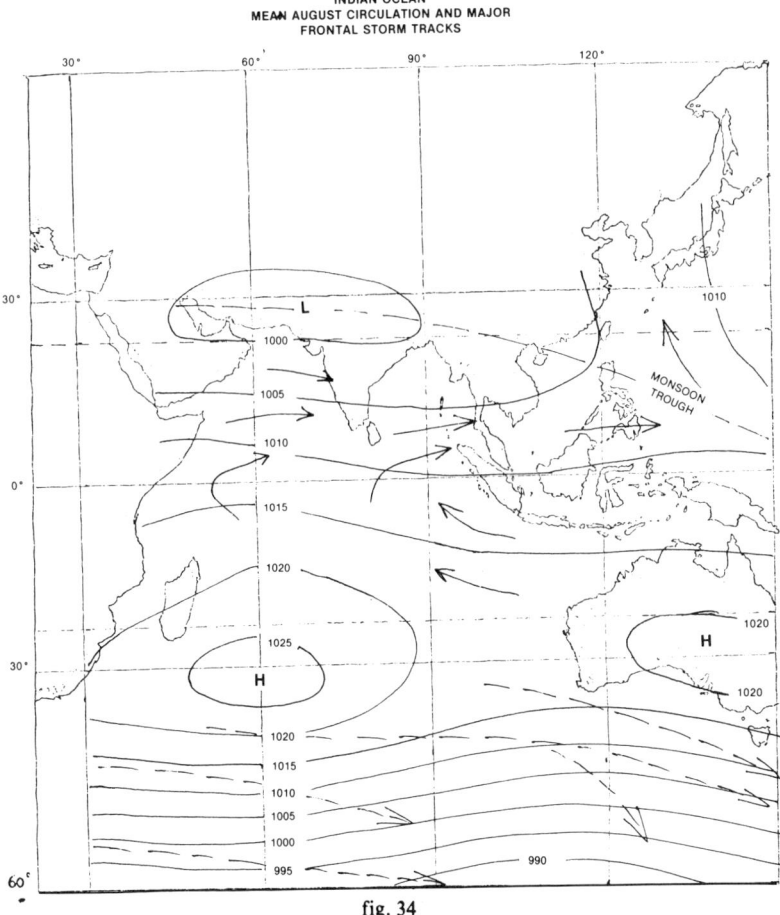

fig. 34

may also form during monsoon "breaks" in summer when the southwesterlies are weak. Most of the storms move west or northwest and enter India or Pakistan, although a few will curve northeast before reaching land. A few rare cyclones reach Saudi Arabia.

Southern Hemisphere

In the south Indian Ocean an average of six tropical cyclones develop with the principal season extending from January to April. Most of the storms form in the western part of the ocean and initially move west or southwest. Some will reach the mainland or islands off

Indian Ocean

east Africa while the rest will curve off to the southeast as they enter the mid-latitude westerlies. An occasional cyclone will also form near the northwest coast of Australia.

At higher latitudes the "roaring forties" storm belt contains frequent frontal storm centers at all months in the longitudes of the Indian Ocean. These storms are usually more intense and their effects may reach lower latitudes during the Southern Hemisphere winter months. (Fig. 35)

3. Indian Ocean Surfing Potential

Arabian Sea and Bay of Bengal Coastlines

Due to the complicated geography and climatology of the Indian Ocean region, it is necessary to make a number of divisions to indicate the best months for surfing. Most of the Arabian Sea and Bay of Bengal will be poor in June through September due to persistent fresh to strong southwest monsoon winds. But a few good days may be possible during monsoon breaks.

The best months are likely to be October and November with a secondary preference for April to early June. This is due to the general light winds and occasional strong swells caused by tropical cyclones. During the Asian winter offshore winds prevail over western India and the parts of southeast Asia bordering the Bay of Bengal. The only probable surf source will be from distant hurricanes in the South Indian Ocean or even more distant frontal storms of the southern oceans. This type of surf should be quite small.

East Coast of South Africa

The surf of South Africa east of the Cape of Good Hope should be consistent and often large most of the year due to the close proximity of the "roaring forties." Cold fronts may cause strong and gusty winds, particularly in late winter and spring (August through October). As one moves northeast from Durban the effects of mid-latitude swells diminish. For the subtropical and tropical east coast of Africa, the best surf should result during the hurricane season of January through April.

South Indian Ocean Islands

There are a number of island groups in the South Indian Ocean that are located in tropical latitudes. These islands should receive swells from both the storms of the Southern Ocean and from Indian Ocean tropical cyclones. This fact indicates that the best chance for surf in

Indian Ocean

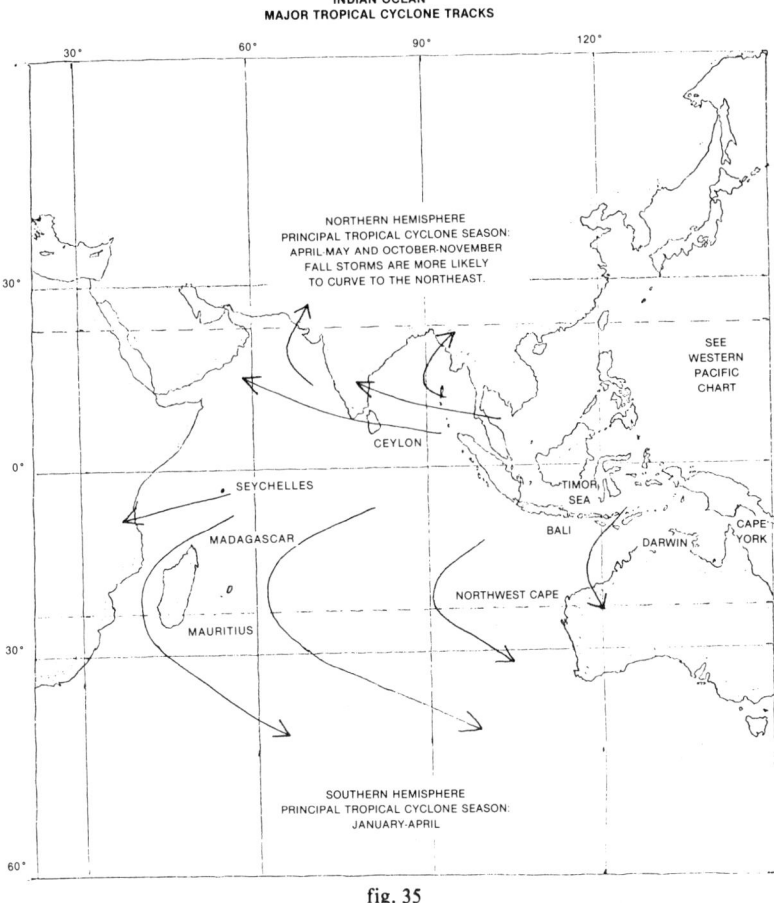

fig. 35

the mid-ocean islands is in summer and again in winter. Surf expeditions to Madagascar, the Seychelles, and Mauritius have found good waves, and other islands await exploration. Recent surfing movies have popularized the large waves found at Bali. These islands have a coastline facing a general southwest direction exposed to swells generated in the monsoon westerly flow near the equator. But, on the mean, the near equatorial westerlies do not reach that far east and the surface winds are southeast. This results in rather consistent and large surf with offshore to sideshore winds in the peak Northern Hemisphere summer monsoon months of July to September. For the rest of the year there will not be much surf, with the exception of possible high waves produced when a hurricane forms near northwest Australia.

Indian Ocean

West Australia

North Coast

Northern Australia from near Darwin to Cape York appears not likely to have much surf, as the adjacent water bodies are too narrow to permit the entry of many swells generated elsewhere. About the only significant wave maker could be a tropical cyclone in the Arafura or Timor Sea.

Northwest Coast

West of Darwin to Northwest Cape the coast may receive distant swells coming from directions of southwest through northwest. Possibly some of the westerly swell reaching Bali could touch the northern-most parts of this area. Infrequent surf could also be expected during the hurricane season, but onshore northwest monsoon flow also is present in these months.

Central West and Southwest Coast

South of Northwest Cape, the western Australian coast is well exposed to swells generated in the higher latitudes. Some surf is likely in nearly all months of the year with the biggest waves likely in the cooler months. In rare cases some surf might come from Indian Ocean hurricanes.

In late spring through fall the prevailing wind blows offshore, with the exception of onshores from sea breezes or frontal passages. Fronts are most common polewards of 30° S.

In winter and early spring southwest Australia is generally quite stormy with frequent frontal passages and strong onshores.

Western Australia is sparsely settled and its numerous points and headlands offer ample opportunity for uncrowded surfing.

South Australia

The southern part of Australia is also exposed at all seasons from swells from the unbroken ocean region surrounding Antarctica. Srong west to southwest swells are likely in the winter months and can occur occasionally even in summer. The wind conditions can be expected to be quite changeable with frequent strong and gusty winds near fronts and storms in the cooler months. A day or two will have offshore winds followed by periods of stormy weather. The waters are chilly and a wet suit is a must most of the year.

Probably the most favorable combination of surf and wind

conditions will be in the spring and fall periods. The surf should be fairly consistent and at times large, with only a few days lost due to adverse winds.

Chapter IV

The Pacific

1. North Pacific Circulation

The world's largest ocean area is the last to be considered. The North Pacific and South Pacific are considered together as a single unit since swells are free to cross the equator largely unimpeded by the presence of major land masses.

Summer

In the Northern Hemisphere the summer circulation is dominated by the Pacific high which is centered on the average near 38° N and 150° W, with a ridge extending westward towards southern Japan. Relatively low pressures prevail in the Bering Sea and moderate frontal storms travel from Japan towards the Aleutians and the Gulf of Alaska. A monsoon trough of low pressure extends from southeast Asia to the Philippines. A pronounced trough forms at times through the Marshall and Caroline Islands near 10° N, and a trough appears south to southwest of Mexico also near 10° N. West to southwest winds will develop between these troughs and the equator.

Fall

During the fall months the Pacific high weakens and shifts towards its winter position near 30° N and 135° W. Frontal storms of increasing intensity and frequency develop east of Japan and move towards the Aleutians and the Gulf of Alaska. A lesser number of storms form north of 35° N in the mid-Pacific and move east to northeast. By November periods of stormy weather and frequent frontal passages reach the U.S. Pacific coast.

In the tropics the near equatorial trough of the western Pacific remains active through November and sometimes into December. The trough of the eastern near equatorial Pacific shifts southward to around 5° N in November. The influence of the near equatorial

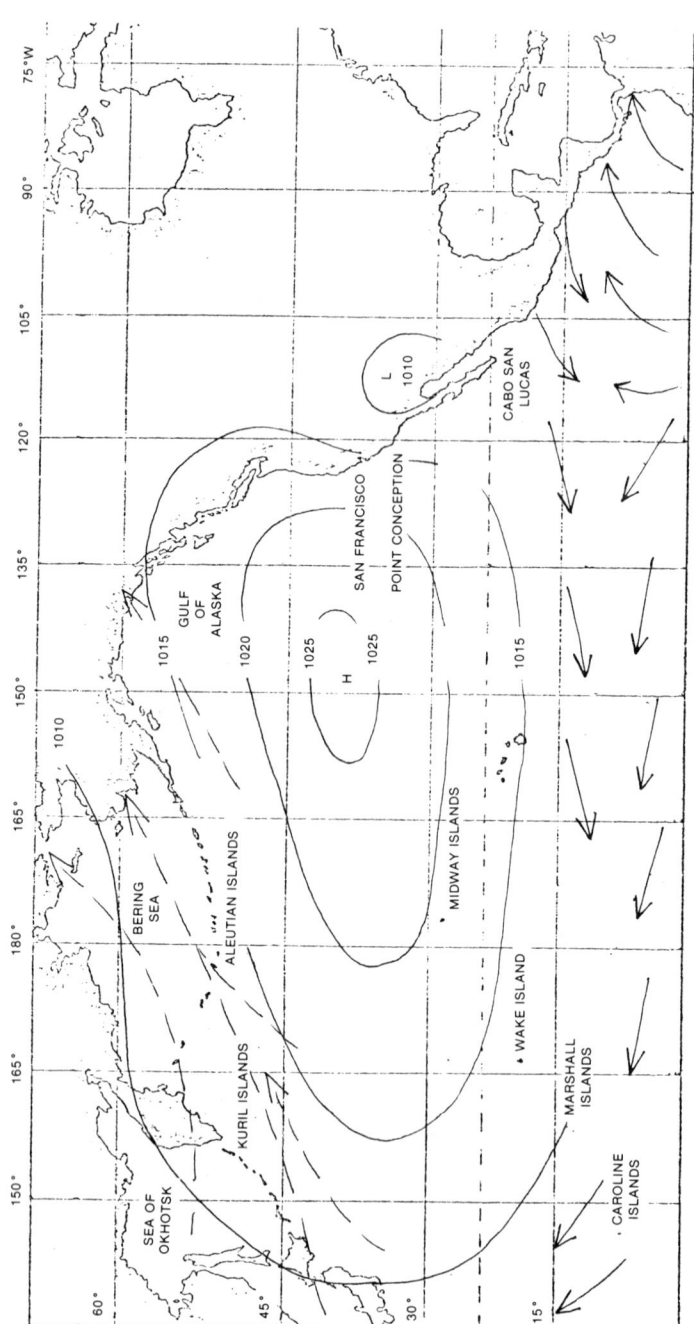

fig. 36

NORTH PACIFIC MEAN AUGUST CIRCULATION
AND MAJOR SUMMER FRONTAL STORM TRACKS

The Pacific

trough position on the length of the Pacific hurricane season can be seen in the next section. (Fig. 36)

Winter

The winter months have a strong Arctic high over Siberia sending blasts of icy northerly winds into low latitudes in the Western Pacific. Frequent and often intense storms follow the primary storm track from south of Japan to the Aleutians and the Gulf of Alaska. Some of the storms can generate near hurricane winds over a thousand miles of ocean lasting for a couple of days. Less common storm tracks originate in the central and eastern Pacific. These storms will move into North America or end up in the Gulf of Alaska. A few low pressure centers that remain nearly stationary for many days called "Kona" storms develop each winter near the Hawaiian Islands.

In winter the near equatorial troughs of the Pacific are located usually near or south of the equator. Steady easterly trades normally prevail south of the Pacific high to the equator across the North Pacific.

Spring

Generally by late March the North Pacific begins a transition back towards more summer-like conditions as the Pacific high intensifies and shifts northwest. The higher latitude storm track becomes less active and fewer fronts reach the U.S. West coast. Usually the near equatorial troughs reach their southernmost position in March or April. By May they will have returned several degrees of latitude into the Northern Hemisphere. (Fig. 37)

2. North Pacific Tropical Cyclones

The North Pacific is the world's most active tropical cyclone producer. An average of 20 to 25 storms per year develop in the Western Pacific. The main typhoon season is long, extending from June to December, and a few typhoons have been reported in all months of the year.

West North Pacific

Most western North Pacific typhoons originate in the area south of 20° N bounded by the Philippines and 170° E. A few also form in the South China Sea. Most storms up through July follow west to northwest tracks quite often reaching the Philippines or mainland

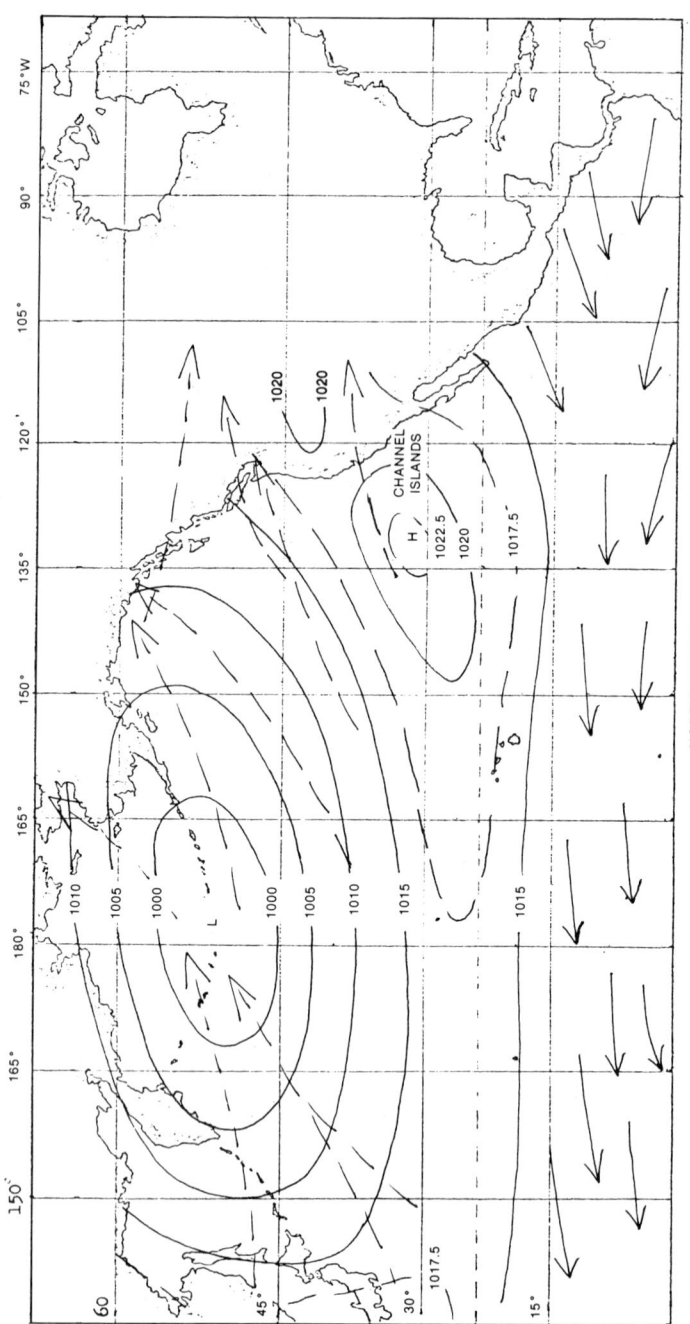

NORTH PACIFIC
MEAN FEBRUARY CIRCULATION AND MAJOR
WINTER FRONTAL STORM TRACKS
fig. 37

The Pacific

China. Occasional typhoons will recurve to the east of Japan.

From August through December about a half of the tropical cyclones continue along west to northwest tracks. The rest curve sharply to the north and northeast, east of Japan. Sometimes the combination of an old typhoon and a developing frontal storm will join together to fuel a very intense higher latitude storm. (Fig. 38)

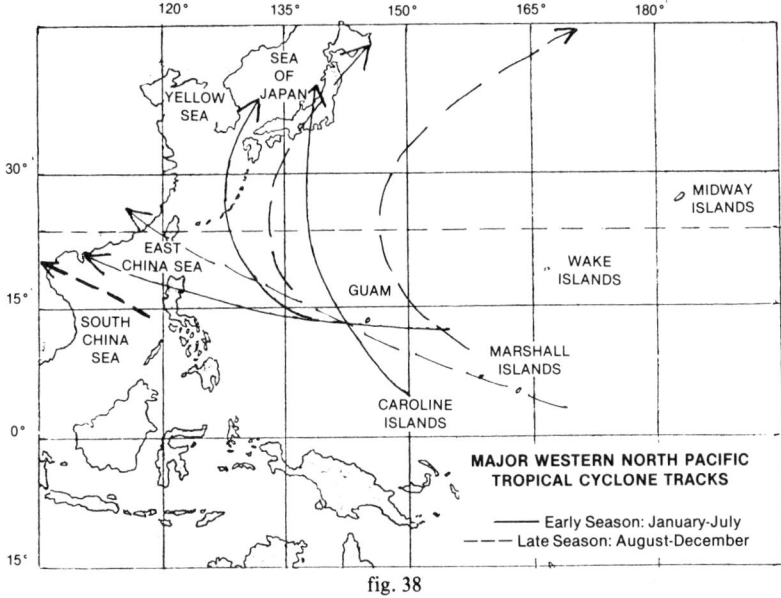

fig. 38

East North Pacific

About 12 to 15 tropical cyclones form from late May to early November in the eastern North Pacific. Most of the storms follow west to northwest tracks away from Mexico and eventually dissipate over the cool waters between California and Hawaii. A few of the storms curve to the north and northeast and will go inland over Mexico. This is most likely after August. About one mid-season (July or August) storm per year follows a long westerly track passing close enough to Hawaii to produce surf there. (Fig. 39)

3. South Pacific Circulation

The South Pacific circultion is driven by the eastern South Pacific high that is found between 25° to 35° S and near 90° W at all seasons. Persistent southeast trades prevail north of the high in the eastern and central South Pacific.

The Pacific

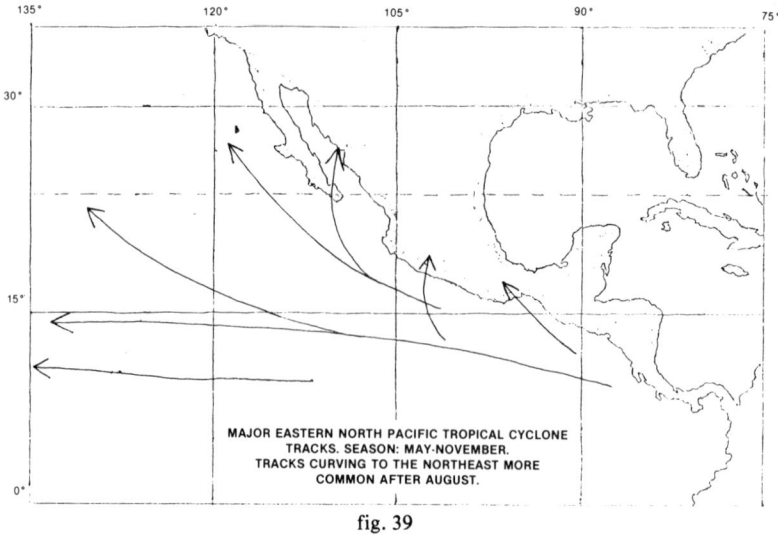

fig. 39

But the wind pattern north of 15° S from Australia to the Gilbert and Ellice Islands tends to be monsoonal. In the Southern Hemisphere summer a trough extends along 10° to 15° S and northwest winds are likely over this area. In winter pressures are high over inland Australia and southeast trades will prevail. Inconsistent shifting winds are likely at other seasons.

Frontal Storms

As in the other Southern Hemisphere oceans intense storm activity occurs at all seasons but is most common in winter over the open water belt between 40° to 60° S. During the cooler months storms may affect lower latitudes, and severe cyclones are sometimes observed near and east of New Zealand. (Fig. 40, 41)

Tropical Cyclones

About eight tropical cyclones form each year in the South Pacific, mostly from December to March. Nearly all the storms originate from 5° to 15° S and west of 180°. In the early stages most of the storms move to the southwest and then recurve to the southeast around 20° to 25° S. Hurricanes move inland occasionally in northeast Australia, and once in a while may reach the north island of New Zealand. New Caledonia, New Hebrides, the Ellice Islands group, and Fiji have

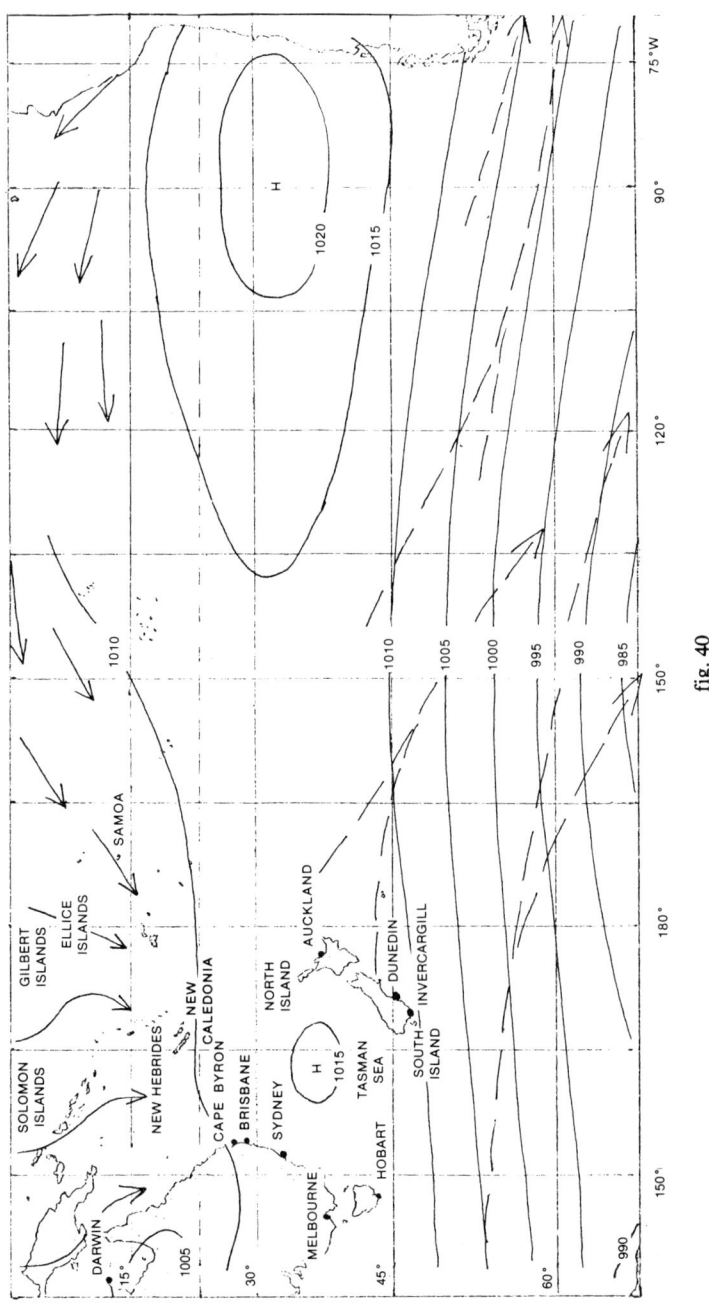

fig. 40

SOUTH PACIFIC
MEAN AUGUST CIRCULATION AND MAJOR WINTER FRONTAL STORM TRACKS

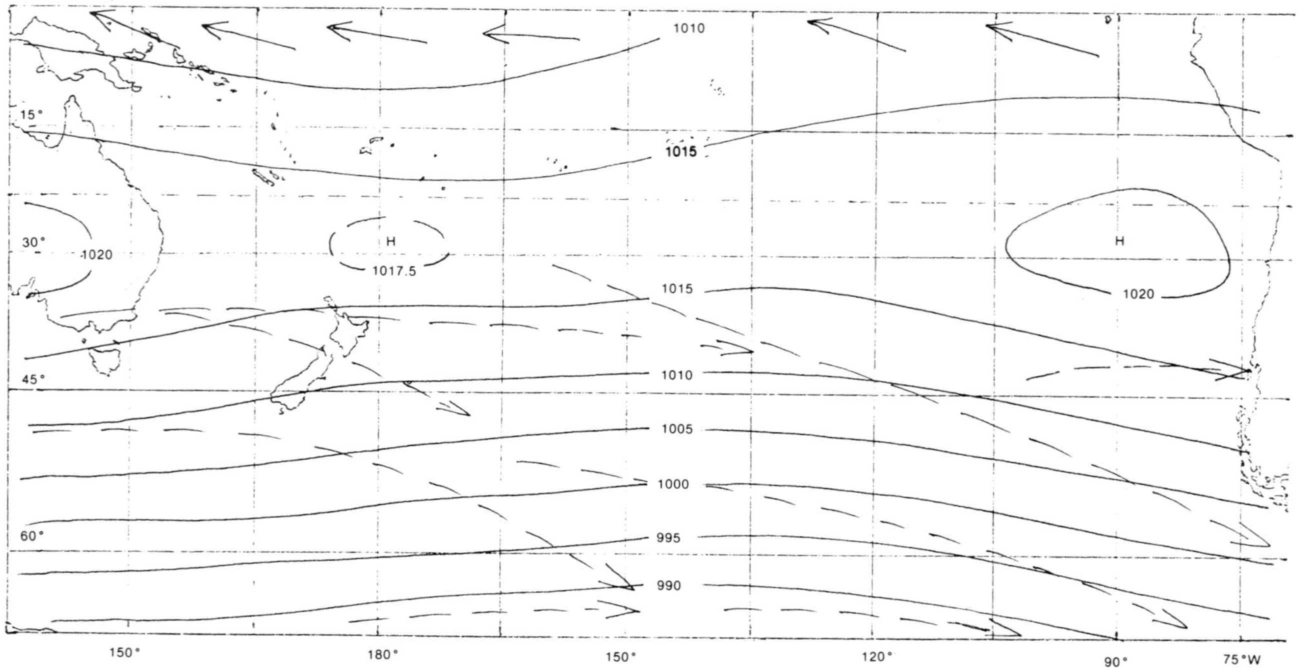

fig. 41

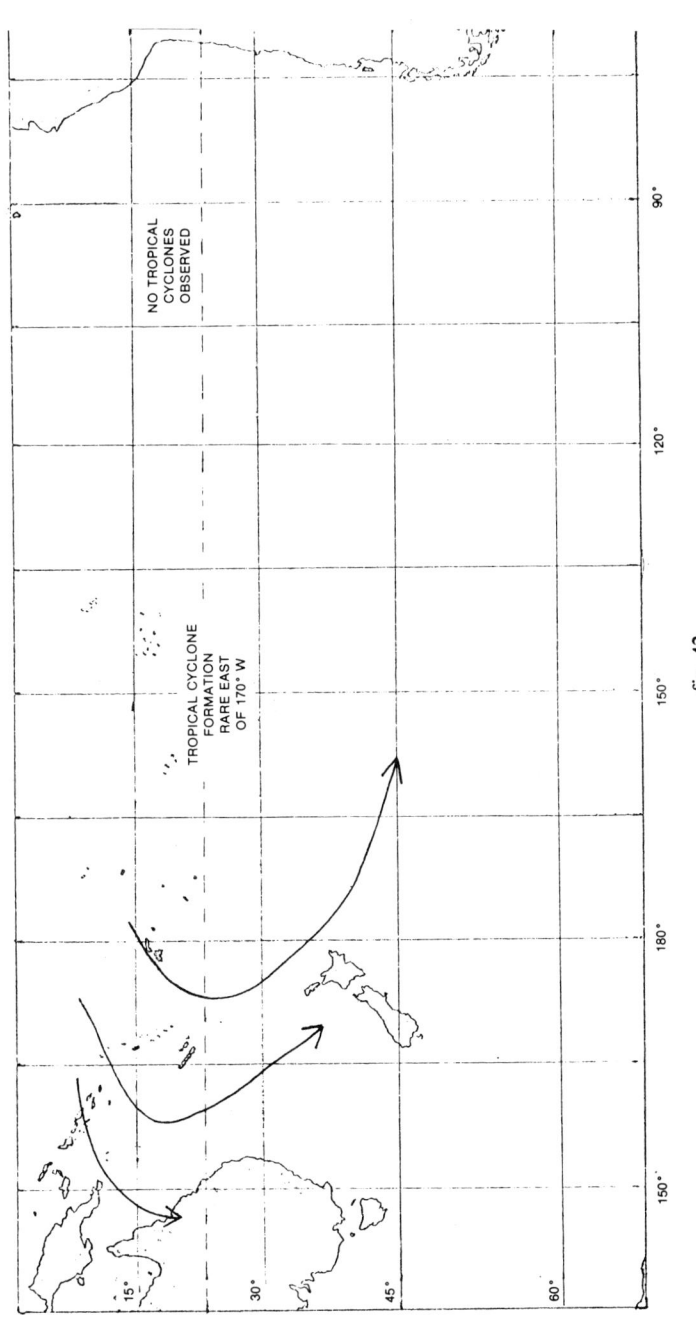

fig. 42

The Pacific

been visited by tropical cyclones. They seldom affect longitudes east of Samoa.

Now we will put this information together and evaluate the surf potential of the Pacific basin. Because of this ocean region's size and complexity, it is again necessary to make a number of subdivisions in the following discussion. (Fig. 42)

4. North Pacific Regional Surf

South China Sea

Although the South China Sea is relatively small, some ridable surf can be expected along its shores at times. During the summer and fall tropical cyclones may enter the region or form there causing swells to radiate through the entire South China Sea. In the winter surges of the northeast monsoon will stir up all or part of the sea. Although these waves are wind-driven, short-period storm surf, they may smooth out and become ridable when the surge ends. This situation may produce surfable waves along the coasts of Vietnam, Malaysia, and a few exposed points in the Philippines.

The Philippines and the north coast of New Guinea likely receive considerable surf from the typhoons of the western North Pacific. In the Philippines westerly winds during the summer monsoon will cause frequent offshore winds, although sea breezes may develop when it is weak. Several typhoons per year pass over the Philippines, so episodes of wild winds and torrential rains must be expected in the primary tropical storm season.

Japan

Farther to the north, Taiwan, the Ryukyus, and the south and east facing coasts of Japan probably have two major surf seasons. In the summer and early fall fairly frequent and sometimes large surf results from typhoons in the adjacent ocean. The water is warm and the winds are usually quite light. The numerous points on the coastline southwest of Tokyo should react well to strong southerly swells.

In winter the ocean is stirred by developing storms east of Japan. High waves will occur for a day or two after a storm has moved away. The wind after a storm passage east of Japan will be offshore but cold and gusty. The water will be icy cold.

Conditions in the other enclosed inland seas such as the East China Sea, Yellow Sea, and Sea of Japan will be similar to that of the South China Sea. Rough seas will produce some short period surf whenever a storm enters the sea. The barrier of islands from the Kurils to Taiwan prevents long period swells entering from the open ocean.

The Pacific

Western North Pacific Islands

The island groups of the western North Pacific such as the Marianas, Carolines, Marshalls, and Wake receive swells from tropical cyclones in the typhoon season and North Pacific frontal storms in the winter months. Unfortunately a number of these atolls are fringed by barrier reefs, making it difficult to reach the surf. Only careful study of topographic maps will reveal likely candidates for surf exploration.

Hawaii

Moving eastward we come to Hawaii, where surfing was born. Hardly a day passes when there's no surf somewhere in the Islands. The north and west facing coasts of the Islands receive their surf from central and west North Pacific frontal storms. Sporadic swells appear in late August and September. By October the surf is quite consistent and often well-formed.

North Shore

Surf increases through November and usually reaches a peak from December through March. Most mid-winter surf ranges in the 6 foot to 12 foot size (measured from the face).

Only a few days will be flat, and several times a winter spectacular 20 foot to 25 foot-waves are likely. Destructive high waves of up to 40 feet or more happened in 1953, 1958, 1969 and 1974.

The mean trade wind conditions are cross shore or slightly offshore at the major surf spots. On some days winds will alternately shift from offshore to onshore, and a close check is needed to keep on top of the situation. And sometimes even when the wind is favorable, swells from two or more sources will combine to produce erratic, shifting, and peaky waves.

After mid-March the swells decrease from the northwest. Several days of small to medium surf will alternate with flat spells. By late May "country" surf becomes rare.

South Shore

At times the south facing coastlines of Hawaii receive swells produced from distant Southern Hemisphere storms. The south swells will cause sporadic surf from April or May through September. This corresponds to the months when the Southern Hemisphere frontal storms are the most intense and closest to the equator. Average summer surf is probably 2 feet to 3 feet, but some 6 foot to 8 foot days occur

The Pacific

each south swell season, and a June 1974 swell hit 15 feet off Ala Moana.

In all seasons persistent trade winds prevail in Hawaii, except for occasional interruptions by storms and fronts in the cooler months. Usually there is an east or northeast swell generated by the trades. This causes very consistent waves at the Makapuu and Sandy Beach body surfing spots on Oahu.

Pacific Northwest

Next we consider the Pacific Northwest as a unit from British Columbia southward to the San Francisco Bay region. Surfers in this region contend with year-round icy water, fog, and frequent onshore winds. There is usually plenty of swell from October to April, but prevailing winds are westerly, directly onshore (except in the lee of pronounced headlands). But offshore winds may occur for a day or so preceding a cold front. This would be the best time to surf.

For the rest of the year the weather is not as stormy, but sea breezes will develop on most sunny days. Usually the surf will be small coming from the west or northwest. Best conditions will be early or late in the day.

Central California

From the San Francisco Bay area to Point Conception the primary swell source will also be from the west or northwest. A few places exposed such as Santa Cruz may also receive swells from a more southerly direction. The surf will be quite consistent and large at times from October or November through the winter months. The best spots will be those having offshore wind conditions following cold fronts. At Santa Cruz the prevailing wind at the south facing coastline is usually side shore. This feature along with heavy kelp beds and a few protruding points makes Santa Cruz the most consistently glassy spot in the central California region.

Southern California

Along the beaches of Southern California, the effects of winter swell will be somewhat less pronounced than they were further north due to the greater swell decay distance. Quite often marked variations in surf height occur over short distances due to the swell blockage by the Channel Islands and by refraction effects over highly varied bottom contours.

Because of its exposure to the southwest this coastline gets swells from the Mexican hurricanes. A few 6 to 10-foot days will happen

The Pacific

each summer, and hurricane swells have produced surf up to 15 feet at Newport Beach. Smaller, very long-period swells also arrive occasionally from the Southern Hemisphere caused by storms as distant as New Zealand.

Waters off Southern California are still chilly, but much warmer than the icy readings of Santa Cruz. The annual range is from about 54° to 72°F. Except for days with cold fronts or Santa Ana conditions, most beaches in Southern California have a regular land and sea breeze cycle. Glassiest conditions are in the first couple of hours of the day or near sunset.

Baja California

The trip to Baja to find uncrowded surf has become popular for California surfers. Conditions in northern Baja are quite similar to those of Southern California. Baja gets some winter northwest swell and surf also some Southern Hemisphere swell. But this area's biggest and juciest waves are usually during the hurricane season. The highest probability of surf is August to October. Once in a while one of the storms may cross the southern part of the area.

The water off northern Baja is quite cold but warms to tropical levels near Cabo San Lucas at the southern tip. Prevailing winds offshore are northwest and land and sea breezes normally develop along the beaches.

South Mexico and Central America

Surfing is increasing along the south facing coastlines of Mexico and Central America. The orientation of the coastline indicates that the major surf producing swell direction is south to southwest.

The largest waves develop from nearby hurricanes. Tropical storm swells are possible from May to November but are most likely from late July through October. During the hurricane season southwesterly monsoon flow prevails between the near equatorial trough (near 10°N) and the equator. At times it becomes strong and persistent enough to cause sizable swells to reach Central America. In addition the months of April through September are also likely times for the arrival of Southern Hemisphere swells.

The winds of the region are north or northeast much of the year, but land and sea breezes are likely at many spots. During the summer and early fall, the southwest monsoon flows may sporadically reach shore, bringing squally weather and ruining wave shape.

Central America has potential for a long surfing season from April to November with the most juice and storms in summer and early fall. Even in the flat months of winter, some surf may come from distant North or South Pacific storms.

The Pacific

5. **South Pacific Regional Surf**

Western South America

While Columbia and Ecuador may get some surf from the Northern Hemisphere summer southwest monsoon circulation south of Mexico, the rest of South America receives most of its swell from South Pacific storms. Tropical cyclones do not form in the southeast Pacific, so there is no surf from that source. The surf should become larger and more consistent from the west and southwest as one moves southward along the coast. Peru and Chile have waves all year, but they are likely to be largest in the cooler months.

Except near Ecuador and Columbia, the ocean is chilly due to the Peru current. Steady southeast trades prevail all year from just south of the equator to 30° to 35°S. The usual coastal fogs and land-sea breeze patterns develop that are expected near cold water.

Polewards of 35°S the winds are erratic due to frequent fronts and storms, and the ocean is icy cold. Probably the best surf sites in South America are in Peru and northern Chile during the winter.

Central South Pacific Islands

Numerous islands dot the western and central South Pacific in tropical latitudes and seem to favor explorations for surf. Nearly all of them should get good sized surf from the south and southwest generated by higher latitude oceanic storms especially during the cooler months. Many of the islands around 20° to 25°S and east of the Dateline should get winter surf as big as the waves during winter in Hawaii. Southern Hemisphere swells have caused 15 foot waves on the south shore of Hawaii, so think how big they were at their source!

Lesser high latitude swell is likely in other months, and once in a while surf will occur during the South Pacific hurricane season of December to March. Also, during these months some northwest swell may appear from North Pacific storms. With two exceptions, most of the area has steady southeast trade winds. The area of the northwest monsoon has already been outlined. During the winter months, it is likely that a few cold fronts might cause shifting winds and some bad weather polewards of 15° to 20°S. The central South Pacific islands area seems to be an ideal region to explore on an extended surf trip especially if one has access to a large sea-worthy sailboat.

New Zealand

New Zealand lies perpendicular to the prevailing westerlies of the 34° to 46°S latitude that the country spans. Ample west to southwest swells will reach the west coast but winds are apt to be frequently onshore and stormy especially in winter. But some fine surf can occur when high pressure settles in, allowing light or offshore winds.

The east coast will generally have much smaller but well-shaped surf due to offshore winds. However, if a winter storm forms to the northeast or east of New Zealand, large waves may reach this coast. Swells approaching from due south will also affect some points of the South Island and southeastern portions of the North Island.

New Zealand water varies from the cool 60's near Auckland to the icy below 50° range in winter near Dunedin and Invercargill on the South Island, so a wet suit is definitely advisable. Considering the wind and swell conditions, summer and fall are probably the best seasons along the west coast. East coast surf is probably the most consistent in winter, although some days will be cold and stormy.

Northeast Australia

The northeast coast of Australia appears to have only one primary surf source, the depressions and storms during the South Pacific hurricane season. Surf is most likely from Darwin to Brisbane in December to March but even then it is apt to be inconsistent. Some swell from the southeast trade wind region will penetrate the Coral Sea, but it is weakened by the Solomon, Santa Cruz, New Hebrides, and New Caledonia island groups. Also the Great Barrier Reef and relatively high incidence of shark attacks reduce the desirability of this region.

Southeast Australia

Farther south from Cape Byron to Sydney and Melbourne the water is chilly in winter. But the coast can get decent surf from refracted south to southwest swells produced in Antarctic waters and sometimes southeast swells arrive from the Tasman Sea. Best months are probably May to October although quite a few days have adverse wind conditions.

In the summer swells are apt to be smaller and inconsistent. But once or twice a year a good swell may be generated by a tropical cyclone a few hundred miles away.

In Conclusion

This ends our look at the world of weather and surf. Remember, the perfect wave seldom comes to one's doorstep. A surfer must travel to find the good, uncrowded, high-energy waves he needs.

Answer the inner call to pass up our plasticized culture and pursue mother nature's gifts. The journey and the destination are well worth the effort.

Go for it!

Good surfing.

—Vic and Joe